Apocalypse

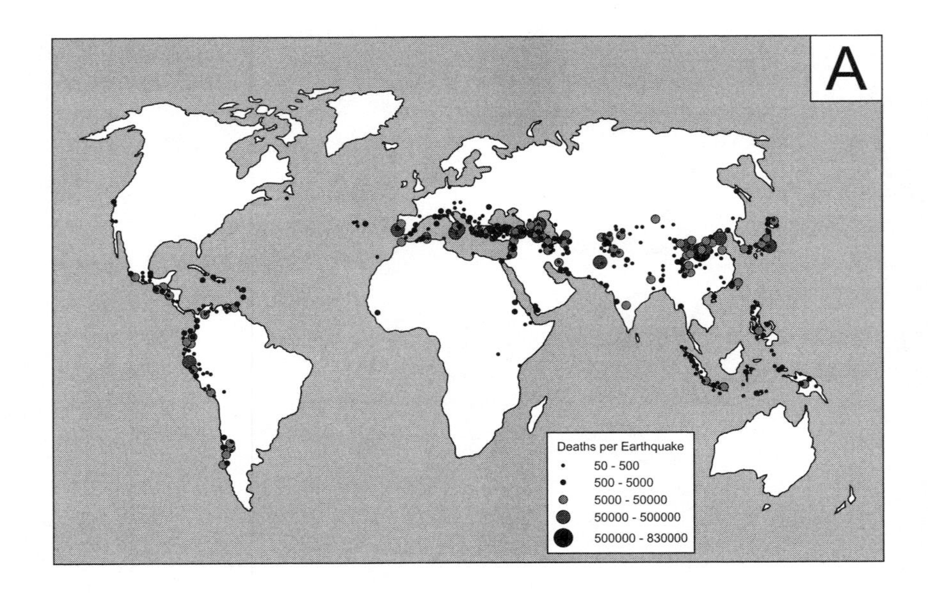

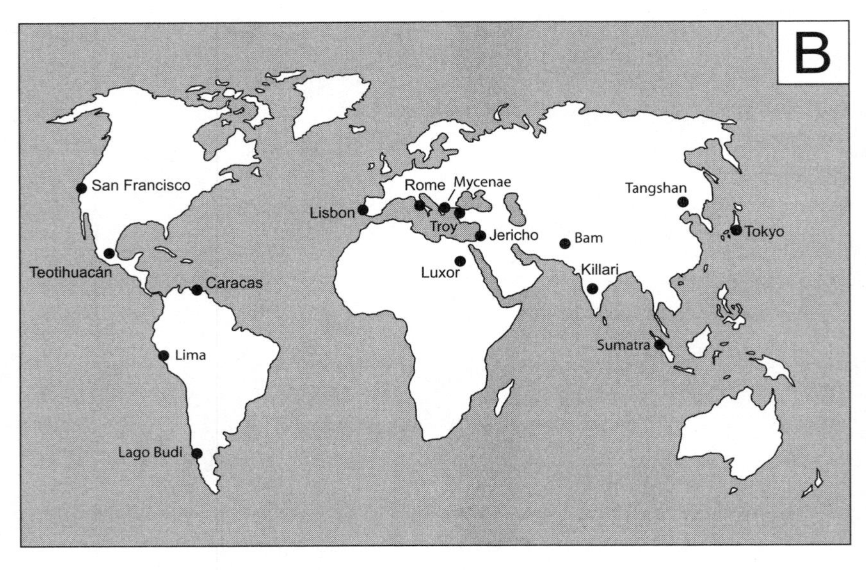

Frontispiece. Geographic relation between earthquakes and archaeology: (a) the most deadly earthquakes in the world between AD 1500 and 2000 (after Agnew 2001); (b) some of the cities and archaeological sites mentioned in this book where earthquakes had a major impact on society.

Apocalypse

EARTHQUAKES, ARCHAEOLOGY, AND THE WRATH OF GOD

Amos Nur

with Dawn Burgess

PRINCETON UNIVERSITY PRESS
PRINCETON AND OXFORD

Published by Princeton University Press, 41 William Street, Princeton, New Jersey 08540
In the United Kingdom: Princeton University Press, 6 Oxford Street, Woodstock, Oxfordshire OX20 1TW

ISBN-13: 978-0-691-01602-3

Library of Congress Cataloging-in-Publication Data
Nur, Amos.
Apocalypse : earthquakes, archaeology, and the wrath of God / Amos Nur with Dawn Burgess.
p. cm.
Includes bibliographical references and index.
ISBN 978-0-691-01602-3 (hardcover : alk. paper)
1. Archaeology and natural disasters. 2. Earthquakes—History. 3. Earthquakes—Social aspects. 4. Earthquakes—Political aspects. 5. Earthquakes—Economic aspects. 6. Excavations (Archaeology) 7. Extinct cities. 8. Civilization, Ancient. 9. Social change—History. 10. Human beings—Effect of environment on—History. I. Burgess, Dawn, 1967– II. Title.
CC77 .N36N87 2008
930.1—dc22 2007026170

British Library Cataloging-in-Publication Data is available

This book has been composed in Sabon with Galahad Regular Display
Printed on acid-free paper. ∞
press.princeton.edu
Printed in the United States of America
1 3 5 7 9 10 8 6 4 2

And they assembled them at the place that in Hebrew is called Armageddon . . . and there came . . . a violent earthquake, such as had not occurred since people were upon the earth, so violent was that earthquake: And the great city was split into three parts, and the cities of the nations fell. . . And every island fled away, and no mountains were to be found.

—Book of Revelation, 16:18–20

CONTENTS

ACKNOWLEDGMENTS

One of the greatest challenges I faced in writing this book was its interdisciplinary—or, more precisely, multidisciplinary—nature. "Interdisciplinary" is a buzzword these days in academia in general, and particularly at Stanford University. It is an appealing idea that is easy to explain but often difficult to execute. A key problem is the need for an investigator or a writer to be able to span several disciplines reliably when he or she is an expert in only one. In my case, I found that branching out from earth sciences (which is my core area of expertise) to archaeology, history, mythology, and social sciences was a risky business. It was not only criticism from others that made me nervous about publishing my ideas but my worry that my knowledge and therefore understanding in those disciplines is incomplete. To compensate for this deficiency—at least in part—I read a lot, and, most important, received encouragement and advice from friends and colleagues outside the earth sciences.

I am indebted to many individuals who, over the past few years, have made it possible for me to write this book. First, I thank those who introduced me to the potential as well as the difficulties of linking earthquakes, archaeology, and mythology to societal and political issues. In 1973, Professor Ari Ben Menahem at the Weizmann Institute of Israel opened my eyes to the richness of archaeological and textual evidence for past earthquakes along the Dead Sea Fault. I also owe a great debt to Professor Nicholas Ambraseys (Imperial College, London). His prolific work is an amazing accomplishment, compiling and evaluating some twenty-five hundred years of textual references to past earthquakes and some of the damage they caused. Even though he was (and, I fear, remains) a staunch skeptic of the value of available archaeological information and a tough critic of what I have tried to develop over the years, I found his work invaluable.

The broadest and deepest contribution to my understanding of the impacts of past earthquakes on history (and, indirectly, archaeology) was from my friend Emanuela Guidoboni from Bologna, Italy. Her monumental works summarizing earthquakes in the history of the Mediterranean region and the Middle East served as a never-ending source of facts and inspiration. I see my book as a modest attempt to bring to the wider public some of the great work that Emanuela did.

Then there are those who helped and collaborated in several aspects of the work that led to my writing this book: Professor Hagai Ron (Hebrew University, Jerusalem) was my early collaborator and partner in exploring the effects of earthquakes on ancient sites in Israel and Jordan. Hagai introduced me to many archaeological sites that I did not know about. Together we produced the documentary *And the Walls Came Tumbling Down*, where we described the impact of past earthquakes in the Holy Land dating back to early biblical times. We also published several papers together on the same topics, including "Armageddon's Earthquakes," which is the basis for parts of chapter 7. Another activity we shared was taking Stanford students on field trips to many of the sites mentioned in this book. Professors Shmulik Marco and Zvi Ben Avraham of Tel Aviv University, Israel, Amotz Agnon, of Hebrew University, and Zeev Reches, now at the University of Oklahoma, were involved in site visits and related work throughout Israel. Professor Renato Funicello of the University of Roma III helped me discover archaeological effects of past earthquakes in Rome and throughout Italy in general. Professor Eric Cline, now at George Washington University in Washington, D.C., published a paper with me entitled "Poseidon's Horses: Plate Tectonics and Earthquake Storms in the Late Bronze Age Aegean and Eastern Mediterranean," which was the basis for chapter 8 of this book.

My colleague at Stanford University, Professor Robert Kovach, introduced me to many sites and facts related to earthquakes in the Americas, including exciting sites in Mexico.

I am also grateful to the many Stanford undergraduate students who, over a five-year period, not only patiently attended my lectures on earthquakes and archaeology but also contributed to this work with their post–field-trip term projects and presentations on sites in Israel, Jordan, Greece, Turkey, Italy, and Mexico. I also thank Lynn Orr, Susan Orr, and Helen Bing for their encouraging support and participation in some of these trips.

Professor Jean Marie Apostolides of Stanford University continuously urged me to write this book, and encouraged me to overcome my interdisciplinary worries.

This book would never have materialized without the dedicated work of my former graduate student, collaborator, researcher, editor, and co-writer Dawn Burgess. Her talent and patience helped transform my scattered writings, class notes, video transcripts, and early drafts into a book. My assistant, Girley Tegama, at Stanford University, has tirelessly sorted out references and illustrations. She has also uncovered leads to materials that were unknown to me. Her dedication to this project has been admirable.

I am also grateful to the earlier help of Margaret Muir at Stanford University, who patiently helped in the initiation of this project, and to Carlo Di Bonito who so generously created the graphics for this book. Carlo helped transform complicated material into clear, understandable visual presentations.

The most sustained support for this book came from Francina Nur, my wife, who has shared with me the discovery of earthquake evidence in field trips to many of the sites referred to in this book. Most important, she tirelessly and patiently encouraged me to complete the project.

Stanford, California, January 22, 2007

Apocalypse

INTRODUCTION

Civilization exists by geological consent, subject to change without notice

—Attributed to Will Durant [US Historian 1885–1981] by Robert Byrne (1988)

The insight displayed by the quote above becomes clear when we combine archeology with earthquake sciences to illuminate the fates of abandoned cities and extinct civilizations. This book was written to explore how earthquakes in the distant past influenced what we have uncovered in archeological sites, and to speculate on the societal, political, and economic repercussions that affected later societies.

Using archeological evidence for the catastrophic, physical collapse of buildings, entire cities, or geographical regions to infer that earthquakes were responsible for the devastation is actually a simple idea but one that yields compelling data. This is especially so in regions where, based on modern geological and seismological data, large earthquakes have repeatedly occurred. It would be ludicrous, for example, to question whether Jericho—2 kilometers from the Dead Sea Fault, the Near East equivalent of California's San Andreas Fault—was destroyed repeatedly by large earthquakes; the question should be which of the earthquakes that struck the area hit when the city was occupied and which when it was abandoned.

Many ruins uncovered by archeological excavations in earthquake-prone regions are the partial result of past earthquakes. The heavy structures of antiquity were designed to support their own vertical weight, but not to withstand the sudden, horizontal ground acceleration that occurs in destructive earthquakes. The Eastern Mediterranean and Near East offer some of the most spectacular examples. Traveling in those parts, one cannot fail to recognize the preponderance of ruins, the many sites that were destroyed and rebuilt again and again. Why are there so many ruins? Is it the result of wars? The passage of time? No, most of this damage is because of earthquakes. The most popular and spectacular sites have succumbed repeatedly to seismic damage: Jericho, Troy, Mycenae, Petra, Knossos, Qumran, Susita, Bet Shean, Jerash, Luxor, and Armageddon, to name a few of the most famous.

We know from modern geological and geophysical research that the Eastern Mediterranean and Near East have experienced a great many earthquakes over hundreds of thousands or even millions of years. The same region has also witnessed, over this vast period, ongoing human settlement and development, and the emergence of great civilizations that created massive structures, including fortifications, palaces, temples, aqueducts, and large masonry bridges. The larger these structures became, however, the more vulnerable they were to damage, if not complete destruction, by sudden earthquakes. The social systems that created these structures may have depended on them for governance and stability, and so the physical destruction of these structures could lead to the collapse of the corresponding social orders. I believe this occasionally happened.

This idea is an example of "catastrophism," the sudden, typically unpredicted natural disaster that leads to abrupt changes in a culture or lifestyle that has been stable for a long time. Following such catastrophes, an entirely new societal, political, or military order can emerge, as seems to have happened when classical Greek culture emerged from its dark ages following the catastrophic collapse at the end of the Bronze Age. Sometimes the only traces of these sudden upheavals are ruins that remind us that what was once prominent, powerful, and stable has suddenly disappeared. For example, the destruction of the Ramesseum and Ramses II statue in the first century BC, as commemorated in Shelley's *Ozymandias,* was surely caused by an earthquake in the Luxor-Thebes area of Central Egypt.

Although these are simple concepts, the idea that earthquakes played an important role in some catastrophic changes in our past—whether in the Eastern Mediterranean and the Near East or in Central and South America—has received stiff opposition. This opposition is, in part, a predictable, professional territorial issue: archaeologists do not want geophysicists to invade their excavations and interpretations, and some historians tend to be skeptical of evidence that is not textual. However, there is a more philosophical aspect that I call the "problem of proof." Mark Rose (1999), the editor of *Archaeology*, had the following response to my paper (Nur 1998), which discussed the role of earthquakes in the cataclysmic end of the Bronze Age:

> It isn't enough to say that the North Anatolian Fault is dangerous and might have unzipped between 1225 and 1175—you need to prove that it did so at that time and, beyond that, show how precisely it would have ended civilization as they knew it, from the immediate effects to ripples through political, economic, and social spheres on local and regional levels.

Rose demanded that, before one can hypothesize that an earthquake destroyed a society, one must prove not only that it happened, but exactly how it happened. Without proof, he claims, such a hypothesis is no more than a Veliskovskyian-style science fiction presented in the guise of science. I believe it is partly as a result of this attitude that some scholars simply ignore earthquakes and other natural disasters, such as volcanic eruptions.

The real question is this: What constitutes proof? The most stringent view (not often held by practicing researchers) insists on a strict interpretation of Karl Popper's notion of falsifiability. In this strict view, no theory or idea qualifies as firm science unless it is possible to devise an experiment that could eliminate the theory if it is false. In other words, evidence supporting an idea, theory, or hypothesis is by itself insufficient to prove its validity. However, in some scientific areas, we do not have the luxury of such strict falsifiability; in geology and archaeology, for example, it is usually impossible to design tests that could falsify a theory. In these disciplines, we have to settle for a much simpler approach based on probability and a preponderance of evidence. This is especially true when we try to predict future drastic system changes or unravel past ones. Can we prove that a major future earthquake will hit the San Francisco Bay area? We cannot. However, the chance that such an earthquake will occur approaches 100 percent, given enough time. Similarly, we could not have predicted the Sumatra earthquake and the disastrous tsunami of 2004, or the Pakistan earthquake in 2005 that left casualties numbering into the tens of thousands. Still, we should have been able to estimate, given past records and our incomplete earth-deformation theories and hypotheses, that such an event would eventually happen.

Can we prove that an earthquake storm ushered in the end of the Bronze Age? Of course, we cannot. We can, however, estimate the likelihood that this could have happened and compare it to that of other alternatives (equally unprovable but even less likely). This reasoning is especially useful for guiding future exploration of sites and for preparing historians, earth scientists, and archaeologists not just to collaborate in the future but also to become reasonably familiar with one another's disciplines.

This necessity—that in some scientific fields we must reason in terms of probability rather than full certainty or proof—has led to Occam's principle:

> Occam's principle states that one should not make more assumptions than the minimum needed. This principle is often called the principle of parsimony. It underlies all scientific modeling and theory building. It admonishes us to choose from a set of otherwise equivalent models of a given phenomenon the simplest one. In any given model, Occam's

> razor helps us to "shave off" those concepts, variables or constructs that are not really needed to explain the phenomenon. By doing that, developing the model will become much easier, and there is less chance of introducing inconsistencies, ambiguities and redundancies. (F. Heylighen 1997)

Although one may read Occam's principle as an excuse for ignorance, it actually represents the most common, widespread practice among scientific researchers.

Some researchers deny that earthquakes, and, by analogy, other sudden natural events, may have played a bigger role in shaping history, simply because these sudden occurrences are not man-made. The temptation of many modern historians, political scientists, and ecologists is to view major disasters in human history as resulting from man's actions. For example, the celebrated historian Toynbee (1939) believed that "the breakdowns of civilizations . . . are not acts of God . . . nor are they the vain repetitions of senseless laws of Nature . . . we cannot legitimately attribute these breakdowns to a loss of command over the environment, either physical or human." It is difficult to imagine a view more diametrically opposed to that of Durant.

Jared Diamond (2005) is consumed with this view in his recent book, *Collapse*. Earthquakes or volcanic eruptions are never mentioned in this book. The cases of cataclysmic breakdown that Diamond includes are all associated with man's actions, not those of nature. Similarly, Tainter (1988) does not consider earthquakes in his extensive review of societal collapses in human history. The ultimate example, however, is the still widely preferred explanation that the catastrophic collapse of the Bronze Age in the Eastern Mediterranean and Near East ca. 1200 BC was a result of invasion by neighboring or far-traveled armies of Sea Peoples or foreign recruited soldiers.

It turns out that these ideas are not based even on Occam's principle. The arguments are circular, proposing that because many of the main centers collapsed into ruins around 1200 BC, the collapse must have been caused by attacking armies. The existence of the ruins is the only proof offered, and human action is blamed by default. With an earthquake storm hypothesis, however, we have at least the potential of a scientifically independent test.

The modern earthquake record indicates that earthquakes occur around the world on planes of weakness in the earth's crust, called faults, and that wherever earthquakes have occurred in the past, they are likely to recur in the future. Where earthquakes are frequent enough, scientists can estimate the probability that future earthquakes will have a certain maximum magnitude and destructiveness, and then plan building codes and public works accordingly. In the many regions where earthquakes have been sparse in modern times, there are often stories or evidence that indicate large earthquakes occurred in the ancient past; in many of these regions, building codes are primitive and rarely enforced, making them vulnerable to the tragic consequences of even small earthquakes.

The new discipline of earthquake archaeology, or "archaeoseismology," brings together the views and tools of archaeologists and earth scientists, in the hope that the combined perspective can extract new information about both the history of society and the risk of future earthquakes. The partnership, however, is an uneasy one, largely because the archaeology community distrusts catastrophism in general and earthquakes in particular as a catastrophic agent. When a city is destroyed for no apparent reason, archaeologists are far more comfortable ascribing the destruction to the vagaries of an unknown enemy than to the whims of nature.

This book reviews the evidence that earthquakes occurred in the past in various archaeological sites, mostly in the Mediterranean region, and correlates the suspected earthquake damage to the known seismic risks of each site. In some cases, there are written records of varying reliability; in others, there is physical evidence that earthquakes occurred; in still others, there is only suggestive evidence and a candidate fault nearby. Every case is controversial, and this book examines both the causes of the controversy and the far-reaching effects of earthquakes on human society.

A SUMMARY OF THE CONTENTS

Chapter 1. King Agamemnon's Capital

At a conference in Mycenae on archaeoseismology, I first grasped the huge gap in understanding and outlook that separates earth scientists from archaeologists. This chapter explores both the archaeological evidence for earthquakes at Mycenae, and the attitudes and preconceptions that shape our interpretations of such evidence.

Chapter 2. How Earthquakes Happen

To understand the signs that earthquakes can leave in the archaeological record, the reader needs to know how and where earthquakes occur. This chapter explains the basics of fault formation, earthquakes, and seismology, and describes how ground motion during earthquakes can damage ancient and modern buildings.

Chapter 3. History, Myth, and the Reliability of the Written Record

Contemporary written records of earthquakes in antiquity are rare, and the strictly historical record is brief; however, many accounts of earthquakes or events that could have been earthquakes are found in the Bible, the *Iliad*, and other pseudo-historical documents. This chapter examines the value of these records as well as how they have influenced archaeologists and scientists.

Chapter 4. Clues to Earthquakes in the Archaeological Record

Earthquakes leave behind many types of deformation, some that are clearly diagnostic of earthquakes and others that are harder to distinguish from the destruction of war or slow decay. This chapter catalogues examples of shifted foundations, fallen walls, deformed arches, widespread fires, and patterns of collapsed columns that could have occurred in earthquakes, and relates them to the seismic environment where they were found. The archaeological sites include Troy, Mycenae, Petra, and many other well-known ancient cities.

Chapter 5. Under the Rubble: Human Casualties of Earthquakes

One of the most telling kinds of earthquake evidence is the discovery of skeletons beneath the debris of collapsed structures. Some critics of archaeoseismology, in fact, point to the lack of skeletal evidence in a site as proof that an earthquake could not have caused destruction there. This chapter catalogues various skeletal finds in famous archaeological sites, some of which have not been widely publicized, and discusses factors, such as the season and time of day when an earthquake hit, that determine the likelihood of finding skeletons in the ruins.

Chapter 6. Qumran and the Dead Sea Scrolls: Destruction That Preserves?

One of the greatest discoveries of archaeology was the Dead Sea Scrolls found in caves in the Judean desert. Many of the caves in the region are filled with rubble that collapsed from the cave ceilings at some unknown time. There are historical accounts of earthquakes in this area and archaeological evidence of earthquake damage at Qumran, which some scholars believe was the home of the scribes who wrote many of the Dead Sea Scrolls. Combining all this evidence leads to a fascinating exploration of how earthquakes may have played a major role in preserving the Dead Sea Scrolls, and how other scrolls may yet await discovery under the rubble.

Chapter 7. Expanding the Earthquake Record in the Holy Land

The goal of archaeoseismology, beyond simply increasing our understanding of the past, is to help seismologists better understand the past pattern of earthquakes around the world, and thereby estimate seismic risks in the future. An accurate assessment of seismic risk is essential for the design of safe buildings and dams. The modern record of instrumentally recorded earthquakes is far too limited to allow us to estimate the seismic risks in many regions, so we must turn to archaeology to help fill in the gaps. This chapter reviews the earthquake record in the Holy Land, and examines how advances in various disciplines are leading to better methods for verifying both archaeological evidence and questionable written evidence for ancient earthquakes.

Chapter 8. Earthquake Storms and the Catastrophic End of the Bronze Age

Large earthquakes can have far-reaching effects on societies, and could, given the right concatenation of factors, lead to catastrophic changes in a region. Of particular interest are sequences of several large earthquakes that occur closely spaced in both geography and time, and can affect a very large region over the span of a few decades. Scholars have proposed that these sequences caused the demise of the Bronze Age civilizations in the Mediterranean region. This chapter compares the modern record of very large earthquakes and earthquake sequences to the areas affected by destruction at the end of the Bronze Age.

Chapter 9. Rumblings and Revolutions: Political Effects of Earthquakes

The most common objection to the hypothesis that earthquakes influenced the end of the Bronze Age is that modern earthquakes do not have lasting effects on society. Although it is true that there has never been a complete societal collapse in response to an earthquake in modern times, the earth's convulsions nevertheless have had major influences on societies when they occurred at times of political or economic stress. We examine some relatively modern examples in Lisbon, Tokyo, and Venezuela.

Chapter 10. Earthquakes and Societal Collapse

It is ironic that, to uncover evidence of past earthquakes, we must overcome the same dismissive attitude toward earthquakes that we are hoping eventually to break down with that evidence. My hope in writing this book is that I can help open the eyes of both the archaeological community and the public to the facts I know to be true: The earth beneath our feet, with its past cataclysms, can be one key to understanding not only our prehistory but our future as well.

CHAPTER 1

King Agamemnon's Capital

Archaeologists of my generation, who attended university in the immediate aftermath of Schaeffer's great work (1948), were brought up to view earthquakes, like religion, as an explanation of archaeological phenomena to be avoided if at all possible.

—Elizabeth French, *Evidence for an Earthquake at Mycenae*

At the entrance to the ruins of the ancient city of Mycenae in Greece, directly beneath the famous Lion Gate, is a sight to make an earthquake scientist stop in awe. The immense stone blocks of the city's outer wall rest atop a smooth, steep incline of whitish, polished rock, a natural bulwark some 4 meters high, which must project unassailable strength to the untrained eye. To a geologist, however, the slick stone surface tells another story. This is a fault scarp, the surface formed by an actively moving fault, where the earth's surface has been violently broken and distorted during earthquakes. The wall on top of the scarp is called a "Cyclopean wall," because its huge, dressed stone blocks are so massive that building such a wall would seem a superhuman feat. Sitting atop the steep slope of the fault scarp, this construction must have reassured the defenders of ancient Mycenae, presenting a formidable barrier to attack. The truth, however, is that the fault itself is a plane of weakness in the earth's crust and is under continual stress. It was a silent, constant threat to the city's oblivious inhabitants (Figure 1.1).

Figure 1.1 In this view of the approach to Mycenae, the light gray surface beneath the Cyclopean wall is a fault scarp, clear evidence of the earthquake hazard at this site.

I first saw this remarkable sight in 1993, when visiting that ancient city with a group of geophysicists and archaeologists, led by Elizabeth French. For decades, Dr. French had been a lead archaeologist excavating Mycenae, and at the time of my visit she was the director of the British School of Archaeology at Athens. As we passed through the Lion Gate, she and the other archaeologists focused on the huge, dressed stones, the impressive work of ancient hands. They did not know the fault scarp for what it was: evidence of the work of ancient earthquakes.

Our tour group had first come together in Athens, as part of an archaeoseismology conference that Dr. French had helped organize. "Archaeoseismology" is a recently coined term for the study of how earthquakes affect archaeology, and this conference was, I believe, the first time a group of archaeologists and geophysicists had ever formally convened to discuss evidence for ancient earthquakes and share ideas across the two disciplines. Two factors, in particular, struck me about the meeting: the first was the beauty of the venue, a restored building directly under the Acropolis. The second was the mutual bafflement that characterized the interaction between the earth scientists and the archaeologists at the meeting.

While we earth scientists presented our seismic hazard maps, geological trenching results, and engineering simulations, most of the archaeologists enjoyed cups of coffee outside as they absorbed the ambiance of the Acropolis. When they presented their results, many of the earth scientists did the same. This inattentiveness, I think, reflected not disrespect between the two groups of scientists, but rather each group's unfamiliarity with the other's methods and jargon. The level of communication, in any case, was not particularly high, but at least the love of Greek coffee gave us some common ground.

Likewise, the tour of Mycenae was something we could all appreciate from our own perspectives. As our tour group sat on the hill overlooking the excavated site of Mycenae, Dr. French explained that it and many of the surrounding towns and villages had been destroyed over a brief period, around 1200 BC. The walls of the city were destroyed in several places, many buildings were completely demolished, and much of the city burned. Apparently, Mycenae was then abandoned for eternity.

Although she was one of the facilitators of the archaeoseismology meeting, and although she agreed that Mycenae had historically been subject to earthquakes, Dr. French remained unconvinced that earthquakes had been responsible for its ultimate demise, still favoring the scenario that invading enemies had destroyed the city.

The attack, she said, was probably part of a larger invasion involving the entire eastern Mediterranean region. Because there were no obvious nearby powers on land, the invaders no doubt came from the sea. Since there were no historical records of the invasions, it was difficult to know for certain where the armies had originated. In fact, this remains a mystery in the field of archaeology. Why did these so-called Sea Peoples suddenly attack? How massive must the armies have been to effect such absolute destruction? Why, after expending the time and resources to overtake whole cities, did they not stay and occupy those places? If they came by sea, how many ships were required to transport the troops, and why have no remains of those ships been found? After bringing the inhabitants to their knees, did they simply load up and cast back out to sea, looking perhaps for another place to pillage? If not, what became of them?

As we followed Dr. French into the city of Mycenae, we pondered these questions. The outer wall fascinated me, with its irregular sections of varying stone sizes and construction styles, each representing the work of a different period in the city's history (Figure 1.2). I was vividly reminded of another important ancient city in my native country of Israel—Jerusalem. Like the walls of Mycenae, the city walls of Jerusalem resemble a masonry quilt, its patches delineating repeated damage and repair through the centuries (Figure 1.3). In Jerusalem, those repairs continue to this day, but in Mycenae, as in many other ancient Mediterranean cities, the repairs ended and the city was abandoned.

In Jerusalem, some patches in the walls are sections that were rebuilt after enemy assaults, but many others were made to repair destruction wrought not by man but by nature. Those sections of

Figure 1.2 This photograph of Mycenae's ruins illustrates at least four distinct styles of wall construction.

Figure 1.3 The ancient city walls of Jerusalem, like the walls of Mycenae, are a patchwork assembly of different construction styles and ages, delineated in this photograph with black lines. Some of these patches are repairs of earthquake damage. Note the incipient crack in the upper left, probably caused by the 1927 Jericho earthquake.

the city walls were toppled by earthquakes, and were subsequently repaired, many times in Jerusalem's past. Confronted with such similar construction patterns in the walls of Mycenae, I had to wonder: Could the collapse of that city around 1200 BC have been caused not by an attacking army from the sea but by an earthquake? Perhaps the city was attacked but by a lesser army, or even by an uprising among the oppressed local populace, and its destruction was then facilitated by an earthquake that compromised its defenses. Either scenario might explain many of the puzzles associated with Mycenae's demise: the suddenness, the absence of a known invader or subsequent occupying force, the immensity of the destruction, and the seeming lack of any strategy whatsoever.

The more I considered this possibility as the tour continued, the more intrigued I became. As the other earth scientists and I explored the site, we found many features that suggested earthquakes. Soon we began to speculate whether earthquakes might also account for the unexplained destruction of other places at the end of the Bronze Age, around 1200 BC. Fallen structures are scattered liberally around Crete, Turkey, Syria, Lebanon, Israel and Jordan—all areas where we know earthquakes are common today. Could the same reasoning also apply in those places?

That reasoning betrayed my geophysics background, and I think it is one factor in the long-standing rift between geophysicists and archaeologists. My immersion in the study of geophysics, of earthquakes and other random natural phenomena, has made me comfortable with a fact that most people have difficulty accepting: that the earth beneath our feet is neither solid nor immovable but moves irregularly and unpredictably. Of course, cataclysmic earth movements occur infrequently, even in the most seismically active areas, but we earth scientists are trained to think not in terms of human years but on much longer, geologic time scales. To geophysicists, a phenomenon that happens every thousand years or so is a frequent occurrence. When we see a fault, we see the inevitability of earthquakes. For us, there is no discussion of *whether*, only of *when*: When did the earth last shake, and when will it shake again?

Archaeologists, on the other hand, immersed as they are in the relics of humankind, in the study of the motives, actions, and consequences of human nature, are preconditioned to seek human action as the cause of human catastrophe. For them, a few thousand years is the largest scope they can hope to encompass within a single paradigm, and even that is a rarity. They also seek to uncover patterns in randomness, patterns that will help them write a history for prehistory. They see the unpredictability of an earthquake as an argument against it, as if, because there is no human causality, it is a *deus ex machina*, constructed to explain the unexplainable, an act of God. Human action is always the preferable explanation.

THE "SEA PEOPLES" HYPOTHESIS

One line of reasoning that archaeologists have worked and reworked in their effort to find a human explanation for the Late Bronze Age destruction at Mycenae and elsewhere is that invaders came by ship to Mycenae, perhaps from Troy in today's western Turkey. Troy itself, however, was destroyed at approximately the same time. Who, then, was responsible for the destruction of Troy? Perhaps it was the Hittites, but it is not at all clear that the Hittites were strong enough to mount a campaign at that time, since their own empire was also collapsing. In fact, around 1200 BC, practically every society in the Mediterranean region appears to have met with major damage or destruction, and, for lack of a better explanation, nearly every instance has been attributed at one time or another to invasion by an unknown enemy from the sea.

The key problem behind the Sea Peoples hypothesis has been the failure (at least so far) to determine the aggressors' identity. The Sea Peoples have been blamed for the collapse of cities in Cyprus, Palestine, Syria, and many others around 1200 BC, but all the likely suspects seem to have been otherwise engaged at the time, either defending their own strongholds or struggling to preserve their own declining social structures. Dr. Robert Morkot (1996) of the University of Exeter writes:

> Despite having been favoured until quite recently, this idea of the Sea Peoples migration can no longer be accepted; there is no real evidence to support it. Essentially the Sea Peoples theory was a convenient and plausible invention of the 19th century, designed [largely by the historian Gaston Maspero] to fit the very limited available facts.

If, however, coordinated attacks by the Sea Peoples were not the cause of all this destruction around 1200 BC, what other explanation is there? One way around the problem has been to assume that the Sea Peoples were actually bands of local raiders and that the destruction of so many sites resulted from general lawlessness at the time. What, then, was the cause of this lawlessness? Many theories have been suggested, from sudden advances in weaponry to climate change. As a geophysicist, I have to throw in my own chip: Could earthquakes have played a part?

One of the great historians of the twentieth century, Arnold J. Toynbee (1939), addressed the causes of societal collapse in his massive work, *A Study of History*. This ten-volume compendium has perhaps had more influence on the thought of historians (and, by extension, archaeologists) than any other work of modern history. In it, Toynbee dismisses the notion that any external influence can be responsible for the collapse of a society. Using examples of some twenty past civilizations around the world, including the Minoans, Mayans, Mycenaeans, Spartans, Andeans, and Hittites, Toynbee argues that, in every case, the cause of societal collapse was internal decay, not external influences of the natural world. (Incidentally, he also dismisses the idea that attacks from without can destroy a civilization, unless that civilization is already on the path to destruction.)

According to Toynbee (1939, 7), "one of the perennial infirmities of human beings is to ascribe their own failure to the operation of forces which are entirely beyond their control and immeasurably wider in range than the compass of human action." He further states:

> We cannot legitimately attribute these breakdowns to a loss of command over the environment, either physical or human. The breakdowns of civilizations are not catastrophes of the same order as famines and floods and fires and shipwrecks and railway accidents; and they are not the equivalent, in the experiences of bodies social, of mortal injuries inflicted in homicidal assaults. (119)

Thus, in a series of negative conclusions, Toynbee asserts that the collapse of human societies is due solely to human failings. He argues that societies follow a progression that, unless certain criteria are met, leads inevitably to collapse from within.

Toynbee's approach presents a curious difficulty for archaeologists. If societal failures are the result of internal social problems, determining the kind of archeological evidence one might find of such an event is difficult. In the absence of a historical record, how does an archeologist find evidence of such internal weakness? This leads, I believe, to an impossible tautology: the only physical evidence of such weakness that the archaeological record can preserve is the fact of societal collapse; ipso facto, the society was weak. In other words, Toynbee's approach does not help archaeologists a bit. It does lead to a corollary approach, however: if the internal structure and weakness of a society always predetermines the society's collapse, then what archaeologists are really looking for are accessories to the crime—triggers, so to speak. This leads again to the predisposition of archeologists to ascribe damage to human causes: a weak or vulnerable society would tend to attract attacks, from without or within the society. Because the strength or weakness of a society would have no effect on the occurrence of earthquakes, of course, earthquakes are not a very satisfying topic for those who study ancient civilizations.

The historian Thomas R. Martin (1996) takes his lead from Toynbee. He discusses several possible candidates for the aggressors who were eventually lumped together as the "Sea Peoples." He adds, however, that internal conflict among the elite, and not necessarily foreign invasion, characterized the destruction of the Mycenaean sites in the period after about 1200 BC. "The destructive

Figure 1.4 Pictured here are four of the countless sites around the Mediterranean where earthquakes played a part in ancient destruction: (a) Jerash (AD 749), (b) Kala'at Namrud (AD 1202), (c) Knidos (AD 460?), and (d) Selinunte (date unknown).

consequences of this conflict," he also points out, "were probably augmented by major earthquakes in this seismically active region."

The controversy over Mycenae's fate highlights an incontrovertible fact about the Mediterranean region: today, from Egypt to Israel, and from Turkey to Greece to Italy, ruined cities and shattered buildings litter the Mediterranean countryside (Figure 1.4). Even given that this region has been inhabited since before the dawn of modern man, why are so many of the ancient buildings and monuments in ruins? Why have countless cities been rebuilt on the rubble of previous construction, only to fall themselves? Because this region is renowned as the cradle of civilization, and of archaeology, we have simply accepted that these ruins are the natural state of civilization's remains and we hardly question the causes of destruction.

Were our ancestors in this region so uniformly destructive that they would consistently reduce their enemies' homes to rubble just to ensure that the structures were no longer habitable? Although scholars (e.g., Drews 1993, 45) often propose this notion as a general modus operandi, they base this assumption only on a few historical accounts of such massive destruction, such as the Roman annihilation of Carthage. If indeed this was the general practice before explosives or modern machinery, it represents an astounding investment of effort and resources. Alternatively, were our ancestors such poor builders that the mere passage of years would cause even their finest stone edifices to topple? I think not. Rather, as this book will show, our modern knowledge of the geography and statistics of earthquakes makes earthquakes the best explanation for many, if not most, of these cases of wholesale destruction.

Even a single earthquake, if severe enough, can cause damage in quite a large region, much larger than is generally accepted in the archaeological literature. Furthermore, today we know that earthquakes on many fault systems occur in sequences, with one large quake after another marching down a major fault, causing damage in a huge region over a just a few years or decades. The episodic destruction of many cities in a given region over a short time would therefore be not a mystery needing an explanation but rather an expected consequence of earthquake damage.

Partially reconstructed ruins around the world draw hundreds of thousands of tourists every year, but the story of how the ruins reached their present state is not always carefully examined. Determining the real story behind the ruins has become even more difficult, as many sites have been reconstructed so that tourists can see how they looked when they were inhabited. Since destruction is almost always attributed to wars and battles, subtle details that indicate earthquake damage are often overlooked, and the act of restoring the ruins erases any evidence that might have remained. In some cases, historical writings tell us about battles that occurred, making such attributions more solid. But in many cases, human action has been proposed simply because of the nature of archaeology: when destruction is discovered, archaeologists are predisposed to look for the action of man rather than nature. The challenge then falls to earth scientists: Can we deduce sufficient evidence from the geography of earthquakes, and from the clues in the ruins themselves, to propose otherwise?

How important is earthquake destruction to understanding archaeology? Archaeologists have not always overlooked earthquakes as the agents that destroyed past civilizations. Indeed, it is the less-than-rigorous invocation of earthquakes by some of their colleagues in the past that has made archaeologists very cautious about proposing such an explanation today. In the early twentieth century, in the layer known as Middle Minoan III of the palace at Knossos in Crete, the British archaeologist Sir Arthur Evans unearthed clear evidence of massive, widespread destruction. The details of his excavation led him to what he considered the best hypothesis: a large earthquake had destroyed the place, probably around 1650 BC (Evans 1928, 1964). In fact, a local earthquake occurred at the very time of the excavation, reinforcing his interpretation.

Many others, however, have cautioned against accepting this conclusion. George Rapp (1986), a geologist at the University of Minnesota in Duluth, well known for his applications of earth science to archaeology, was one of these critics. Rapp argued that Evans should have used multiple hypotheses to account for the destruction, rather than settling on an earthquake as the most probable cause.

Although I agree that multiple hypotheses are always useful, archaeology has its own tradition of how theories should be set forth in the final report of an excavation. In archaeological literature, the excavator traditionally proposes what he considers the most likely scenario, perhaps also mentioning other probable scenarios when there is conflicting evidence. Had Evans proposed invasion and sacking as the cause of the destruction at Knossos, his assumption would probably have gone unchallenged, with the argument centering instead on the identity of the invading party. Perhaps an earthquake hypothesis is more likely to be challenged than an invasion, simply because the former explanation leaves no room for corollary arguments about identities and intentions.

Evans was more justified than he knew in proposing earthquake damage at Knossos. Although his own experience and reports acknowledged that the area was subject to earthquakes, he did not have the data available to us today (and to Rapp, for that matter) through modern geophysics. The island of Crete is actually the locus of some of the largest and most frequent earthquakes in the entire Mediterranean Basin. Evidence from plate tectonics, discussed in more detail in subsequent chapters, tells us that the continent of Africa is slowly diving under Europe, and that Crete, in the collision zone, must have been devastated many times since the first Minoan Palace was built at Knossos more than four thousand years ago. Evans's hypothesis was consistent with this pattern. Not only was he correct that an earthquake was the most probable cause for the destruction of his Knossos palace, but similar earthquakes and devastation must have happened many times before and since. In fact, the Minoan stratigraphy, with its nine distinct periods, may represent, at least in part, that history of repeated, large earthquakes.

I argue that the non-archaeological evidence for frequent earthquakes in Crete is irrefutable. Geophysical evidence alone leaves little doubt that the entire island has been subjected to repeated devastation in the last few thousand years. That alone should make earthquakes not the explanation of last resort but rather one of the first suspicions to come to mind when we see widespread destruction in an archaeological site in that region.

I understand archaeologists' caution: just because earthquakes are common in certain areas does not mean that they lurk under every fallen column or collapsed wall. Although my expertise is in geophysics, I also recognize that we must not ignore the rich and diverse body of archaeological literature. Certainly, the seismic activity of the Mediterranean in no way diminishes the cultural and political complexity of the area. However, whenever I initiate a discussion of prehistoric earthquakes with an archaeologist, the response, nearly every time, is uneasy skepticism.

An interview I gave in 1994 brought this reaction to the forefront. Tom Naughton, a producer for The Learning Channel on TV, was running behind his production schedule when he called me to ask if I would participate in a half-hour documentary about archaeology in Israel. I had already been planning to visit Israel the following week, so I accepted his invitation to meet with the production crew there. The documentary was to be about Dor and Megiddo, two ancient towns in Israel that had experienced mysterious destruction around 1000 BC, which I will revisit later in this book. I talked about the geophysical evidence for earthquakes in this region, gave my opinion that earthquakes were a likely cause of the destruction (given physical evidence uncovered at the sites) and left to await production of the program.

Months later, I watched the completed documentary, which had the wonderfully sensational title *Killer Quakes of the Bible* (Rhys-Davies, 1994). To my surprise, in the finished product I participated in a debate with archaeology professor Amnon Ben Tor from Hebrew University in Jerusalem, whom I had never actually met. However, I knew of him through his father, Yaccov Ben Tor, who had been one of my favorite geology professors when I was an undergraduate student at Hebrew University.

I watched myself explaining that earthquakes must have struck at Megiddo and Dor repeatedly in the past, and that one of the earthquakes could have caused the destruction discussed in the documentary. Amnon Ben Tor responded, "It is a nice story, a nice interpretation, a nice possibility. But to say that we are 100 percent certain—I don't think we can."

Of course, Ben Tor is correct in some sense. When pressed, few scientists will say that they are 100 percent certain of anything, and I am no exception. Certainly, geologists and archaeologists have this in common: they can rarely *prove* their hypotheses. In fact, Karl Popper, one of the twentieth century's most influential philosophers of science, asserted that this is a litmus test for science itself, that there is no way to prove a scientific hypothesis *true*; we can only prove that an idea is *false* when it is contradicted by evidence. The best scientific ideas, according to Popper, were those that made predictions that could be proven false if the ideas behind them were incorrect. It is only the repeated failure to prove predictions *false* that gives us some confidence in the relative accuracy of a theory (Popper 2002 [1959], 1974).

In this respect, geology and archaeology are in a worse situation than many other branches of scientific pursuit, for one simple reason. They are not, and never can be, experimental sciences. We cannot turn back the hands of time to observe how a rock or an ancient clay vase was formed or destroyed. We cannot keep notes in our lab books, rewinding and repeating the history of a site until we are confident we have eliminated all possible sources of error or bias. Instead, in both sciences we end up doing detective work—assessing plausibility, probability, and internal consistency. We can use the available evidence to eliminate scenarios that are not consistent with some of the data, and we can examine the possibilities that remain as dispassionately as possible, trying to find tests that will distinguish among them.

There is a scientific trap, however, that we must be careful to avoid, one to which our training as scientists makes us particularly prone. The experimental sciences teach us that good experimental design involves isolating our hypothesis from other possible influences, and changing only one variable at a time. Thus, we take great care to ensure that no competing effects are present that may mask the interaction we wish to study. Nature and history, however, are not so careful. One can never assume that the scenario we wish to test—invasion, revolution, economic collapse, famine, or earthquake—happened in isolation. This makes it much more difficult to decipher archaeological clues.

In an ideal situation, it should be possible to make falsifiable predictions based on the hypothesis that an earthquake struck a given region. We cannot say, based on architectural remains, whether a given building would have survived an earthquake, but we can use local geology to predict which sites were particularly susceptible to damage in a given earthquake and which had soil structures or bedrock that afforded greater protection. We do this today, making maps to predict the intensity of seismic shaking in populated regions as a way to determine modern earthquake hazards. If we wish to test such "predictions" in the archaeological record, to see whether damage patterns in ruins match our risk estimates, it is crucial that the excavators in the predicted region of damage be aware of the hypothesis and that they specifically look for the presence or absence of signs of earthquake damage. For this to happen, we must remove the stigma that the word "earthquake" carries in the archaeological community.

The intent of *Killer Quakes of the Bible* was to shed light on the ruins at Megiddo and Dor, and to produce an interesting story about the always-popular topic of biblical earthquakes. An unexpected benefit was that the documentary also brought archaeologists and earth scientists together to look at each other and at the scientific ground that they share.

Even after the documentary, however, I still did not fully understand why archaeologists seemed so uncomfortable with earthquake scenarios. For example, on our Mycenae field trip, even surrounded by a crowd of geologists pointing at the fault scarp, which still exhibits the polished surface that indicates geologically recent activity, Elizabeth French remained quite skeptical that an earthquake might have destroyed Mycenae. Later, however, with more time to think about the evidence, she wrote a brief paper in 1996, "Evidence for an Earthquake at Mycenae," cautiously acknowledging the possibility. Even then, earthquakes were such a touchy subject for her that, throughout the paper, she never referred to an earthquake without enclosing the word in quotation marks. For me, though, the most enlightening part of the paper was an unusual sentence in the abstract, which at last made clear to me why most archaeologists conducting excavations since the mid-nineteenth century have been so reluctant to consider earthquakes as a destructive agent.

"Archaeologists of my generation," wrote French, "who attended university in the immediate aftermath of Schaeffer's great work, were brought up to view earthquakes, like religion, as an explanation of archaeological phenomena to be avoided if at all possible." The "great work" to which she was referring was *Stratigraphie Comparée et Chronologie de l'Asie Occidentale (Comparative Stratig-*

raphy and Chronology of Western Asia) by Claude F. A. Schaeffer, who at the time of its publication, in 1948, was the research director of France's National Center of Scientific Research, and conservator of the French National Museum of Antiquities. In this work, Schaeffer noted strong changes in cultural patterns in the stratigraphy at sites throughout Anatolia, Syria, and Palestine, dating to between 2400 and 2300 BC, and, most significant, to around 1200 BC. Based on a careful compilation of results from many individual excavations, and his own work at Ugarit (in today's Syria), Schaeffer concluded that the cause of this last catastrophic societal change, which defined the end of the Bronze Age, was a great earthquake. Schaeffer's map (Figure 1.5) details sites for which he proposed Late Bronze Age earthquake destruction.

The reaction to Schaeffer's hypothesis was immediate and sharp criticism, bordering on ridicule, and, in the end, his "great work" was somewhat of a blow to his career. The main arguments against his hypothesis were that the earthquake required for such a catastrophe was unacceptably large, and that there was no evidence that such large events could happen in that region. Persistent ridicule from his peers (e.g., Hanfmann 1951, 1952; Ambraseys 1971; and Rapp 1986, 371, 37) greatly diminished Schaeffer's reputation, and his example served as a caution to his colleagues and successors, such as French.

Fear of ridicule is a great shaper of scientific orthodoxy, and Schaeffer probably suffered a great deal because of the scorn he faced. By the end of this book, however, it will become clear that, although Schaeffer's proposal cannot strictly be proved true, it was not at all a ridiculous idea, at least not for geological reasons. Unfortunately for him, the earthquake record he had at his disposal in 1948 was extremely poor—much shorter and less detailed than what we have today—and scientific understanding of earthquakes was still rudimentary, even among seismologists. In this book, I examine some of the evidence we now have for current seismicity and earthquake activity in some archaeologically important regions. I discuss ways that earthquakes damage buildings and cities, and how ancient earthquake damage would compare to what we

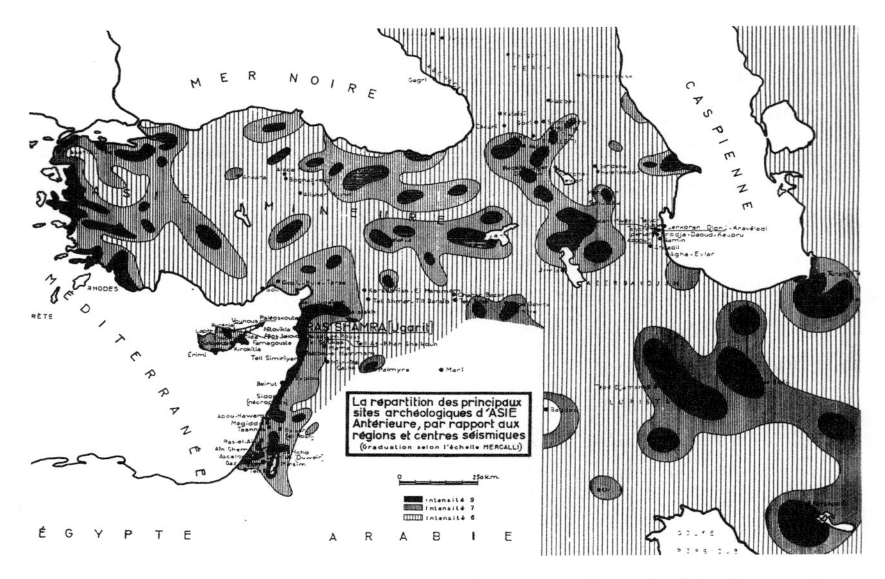

Figure 1.5 Sites with possible earthquake damage at the end of the Bronze Age (Schaeffer 1948). The dark areas are those with significant damage.

see around the world in modern earthquakes. I also discuss areas where I think earthquakes have had a significant impact on sites of archaeological interest, particularly in the Mediterranean region, including Greece, Turkey, and the Holy Land.

Archaeologists and seismologists have a great deal to learn from each other. Other fields in earth science have already become integral to the archaeological method. One example is palynology, the study of fossil pollen, which allows archaeologists to examine how climate and agriculture relate to the remains of structures and tombs. Archaeologists may embrace palynology partly because it, like other areas of geological stratigraphy, uses methods already familiar to them. Conceptually, a fossil pollen grain is no different from a scarab or a coin; they are independent, datable objects catalogued in different layers of an excavation site.

Archaeoseismology is another story. Earthquakes leave no telltale particles that we can bag and catalogue. Rather, we must deduce the evidence by analyzing the geometrical relationships between artifacts, ruins, and human remains, and by analyzing regional patterns

Archaeological Periods in the Middle East

Archaeological literature often dates objects and layers in excavated sites by archaeological period instead of giving a chronological age in years. The name of each archaeological period derives from the predominant technology or culture of the time, so that technological or political revolutions in the archaeological record are accompanied by a change in period name. The progression of archaeological periods in an area is determined by correlating excavation layers within one site and among sites within the region, until a relatively complete coverage is obtained back through time. Rather than establishing an absolute calendar date for an object or event, an archaeologist determines a relative date, bracketing the object in time between other objects or events already catalogued.

There are many advantages to this method; most important, the names of these periods can remain unchanged, even when their absolute dates are changed. Historical records, independently dated objects like coins, and radiometric dates can all be used to pin the archaeological time scale to absolute calendar dates, which sometimes are much earlier or later than archaeologists had thought. If previous literature had specified precise calendar dates, then all of that literature would have to be revised to include the new data.

The archaeological timescale below lists approximate absolute dates for the most commonly cited archeological periods. Note that these dates are estimates of when these periods first appeared, generally in the Middle East; the Bronze Age in Britain, for example, would have been hundreds of years later than in the Middle East, as technology was very slow to expand to some frontiers. Note also that there are at least as many controversies about these dates as there are archaeological periods. These dates are the ones most generally accepted by the majority of biblical archaeologists.

Paleolithic (Old Stone Age)	before 10,000 BC
Mesolithic (Middle Stone Age)	10,000 to 8000 BC

continued on page 30

Neolithic (New Stone Age)	
Prepottery	8000 to 5500 BC
Pottery	5500 to 4000 BC
Chalcolithic (Copper Age)	4000 to 3000 BC
Bronze Age	
Early Bronze Age	
EB I	3000 to 2800 BC
EB II	2800 to 2500 BC
EB III	2500 to 2200 BC
EB IV	2200 to 2000 BC
Middle Bronze Age	
MB I	2000 to 1800 BC
MB II	1800 to 1500 BC
Late Bronze Age	
LB I	1500 to 1400 BC
LB II	1400 to 1200 BC
Iron Age	
Iron I	1200 to 1000 BC
Iron II	1000 to 600 BC
Babylonian and Persian Periods	586 to 332 BC
Hellenistic Period	332 to 37 BC
Roman Period	37 BC to AD 325

of destruction. All this analysis relies on statistics, and only in rare instances can we determine with real certainty that an earthquake, and not human action, was responsible for destruction. We have much to gain, however, by trying to come to terms with this uncertainty and to include earthquakes in the list of legitimate subjects of archaeological investigation.

By learning about earthquake damage patterns and how earthquakes occur, archaeologists can overcome their misconceptions about ground displacement and epicenters, and develop a better understanding of the kind of evidence that constitutes a likely indicator of earthquake damage. By incorporating archaeological evidence, geophysicists may be able to close gaps in the earthquake record and get information about previously unknown earthquake hazard areas. For this to happen, however, investigators from both disciplines must coordinate their efforts and know what to look for.

Although even today seismologists are unable to predict precisely when and where a particular earthquake will strike, they do have information about the general distribution of earthquakes. Data collected with modern instruments in the last hundred years or so have revealed a great deal, not only about where earthquakes occur but also about *how* they occur, and how their damage is distributed. Armed with this knowledge, we can set out to connect the physical geography of a region to the prevalence of earthquake destruction there, and archaeologists can begin to do more than guess at what an earthquake looks like a thousand years later.

CHAPTER 2

How Earthquakes Happen

> A bad earthquake at once destroys our oldest associations: the earth, the very emblem of solidity, has moved beneath our feet like a thin crust over a fluid; one second of time has created in the mind a strange idea of insecurity, which hours of reflection could not have produced.
>
> —Charles Darwin, Journal of researches into the natural history and geology of the countries visited during the voyage of *HMS Beagle* round the world, 1845

In the words of Grove Karl Gilbert, one of the most celebrated American geologists, "It is the natural and legitimate ambition of a properly constituted geologist to see a glacier, witness an eruption and feel an earthquake" (Gilbert 1906). On October 17, 1989, at 5:04 PM, I had my own chance to fulfill the third part of this ambition when the Loma Prieta earthquake hit Stanford, along with the rest of north-central California. I ducked under the desk in my office, just as my steel bookcases crashed onto the chair where I had been sitting (Figure 2.1). Somehow, I escaped injury.

The Loma Prieta earthquake, centered in the Santa Cruz mountains, was no surprise to geophysicists, since it is well known that the entire California coast is subject to repeated major earthquakes along the San Andreas Fault zone. Yet, even those of us who should know better can become complacent, neglecting basic earthquake-preparedness measures like securing tall bookcases to the walls.

Figure 2.1 Photograph of my office at Stanford University, immediately after the 1989 Loma Prieta earthquake. I was sitting in the chair indicated by the arrow when the earthquake struck; by diving under the desk until the shaking stopped, I escaped injury.

Earthquakes have accompanied people since the beginning of human experience. The Loma Prieta earthquake was just one of the estimated several million that occur annually around the world. Most of these elicit little reaction, either because they strike remote, inaccessible areas or because they are too small to feel and can be detected only by very sensitive seismic instruments. The United States Geological Survey (USGS) alone records almost twenty thousand earthquakes per year, most too small to cause even local damage.

However, a few earthquakes—the large ones—are terribly destructive. Even before humans built structures that could fall on them, large earthquakes have had deadly consequences. The oldest evidence for this is found in the Zagros Mountains of northern Iraq, at one of the most unusual archaeological sites in the Middle East. The Shanidar Cave (Figure 2.2) is a huge, natural limestone cave 175 feet wide and 130 feet deep, with a ceiling 45 feet high. Partial excavation of its packed-dirt floor has yielded evidence of

Figure 2.2 The entrance to the Shanidar Cave in Northern Iraq. This cave, inhabited more or less continuously for more than one hundred thousand years, suffered several ceiling collapses during human habitation (Ralph S. Solecki, Columbia University).

more than a hundred thousand years of nearly continuous habitation by modern humans and their forebears. In fact, Kurdish goatherds were still living in the cave when Professor Ralph Solecki of Columbia University began excavating it in 1951.

Solecki (1959), digging down 45 feet to bedrock over four field seasons, made many exciting discoveries in the Shanidar Cave (Figure 2.3). Artifacts and animal bones comprised a large part of his find, along with a graveyard containing twenty-eight anatomically modern humans. The most unusual finds by far, however, were the eight Neanderthal skeletons he unearthed. Most intriguing to archaeologists and anthropologists are a few Neanderthal graves

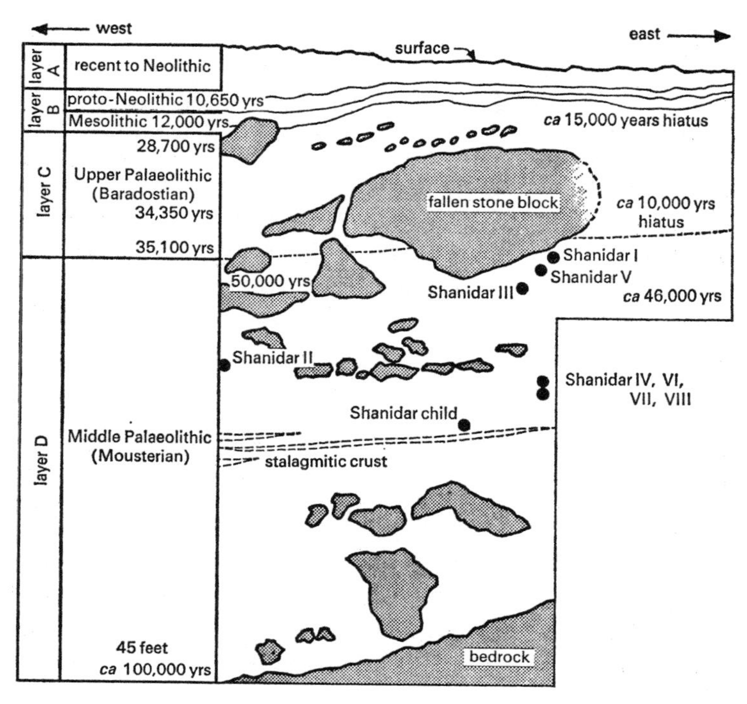

Figure 2.3 Cross-section of Solecki's excavation of the Shanidar Cave floor. The dots represent Neanderthal skeletons, some of which seemed to have been buried deliberately; others were apparently killed by falling rocks from the cave's ceiling (Ralph S. Solecki, Columbia University).

inside the cave, where Solecki found evidence (from fossil pollen) that large quantities of flowers had been buried with the dead. This discovery implied a tender, emotional side of Neanderthal man that no other excavation had ever revealed. As a geophysicist, though, I am more interested in the remains of those unlucky individuals who were not accorded deliberate burial by their fellows.

Stone blocks falling from the cave ceiling apparently killed several of the Neanderthals Solecki found at Shanidar. From the distribution of large boulders mingled with the layers of detritus, Solecki estimated that four major ceiling collapses occurred during the human occupation of the cave, the first three found at depths of 30 feet, 24 feet, and 16 feet below the modern surface, and the fourth found near the modern surface. There were also at least twenty other, more minor rockfalls. Solecki immediately attributed the rockfalls to earthquake damage, noting that a fault scarp—not unlike the one at Mycenae—was apparent near the entrance to the cave. He did not give the scarp much personal attention, however, until the last few days of his second field season in the cave. While cleaning up and making a final record of the season's progress, Solecki himself experienced an earthquake in the cave. His own words, excerpted from his book, bring to life the shock of the experience:

> On August 14 we had a small crew of men to help in the final operations at the excavation. We had dug down to a depth as great as the ceiling was high above the floor of the cave, or 45 feet. I made three detailed profile drawings of the excavation. . . . To make the record complete, time-exposure photographs were taken of all the walls. It was while I was engaged in this task at 2:10 PM on this brilliant day that I almost joined the Neanderthals with a couple of faithful workmen. I had set up my camera on its tripod at the very base of the pit, and was fussing with the lens adjustments, when there sounded something like a tremendous clap of thunder. . . . At the same time, some dirt and dust came down on me and my camera, much to my annoyance. Glancing up to see what could have caused the commotion and hubbub above, I was on the point of bawling someone out for getting too close to the edge of the pit and fouling up my camera view finder. It was only then that I found that the sky, seen as a patch of blue out of the excavation, bore not the slightest trace of my expected storm cloud. And then I suddenly realized the unusual nature of the disturbance. One of my workmen indicated to me that the rock he had been sitting on had shaken. It had been an earthquake. We lost no time in scrambling up the ladders lest the ceiling come down on us, crushing us like flies, or as in the Shanidar situation (as we discovered the next season), like Neanderthals. (Solecki 1971)

The skeletons found in the cave, including those that appeared to have been deliberately buried there, were compressed, broken,

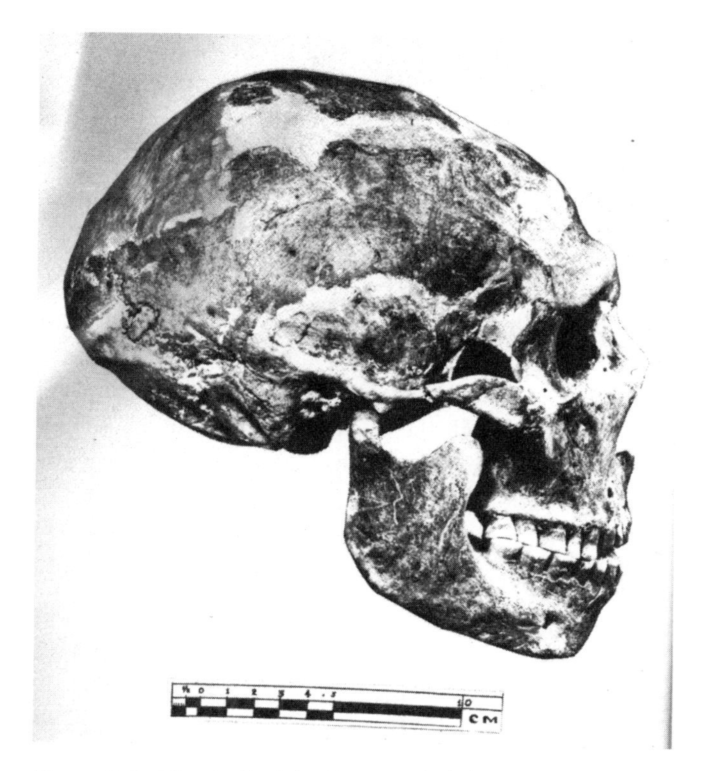

Figure 2.4. The skull of the Neanderthal designated Shanidar I. Details of the breakage pattern of this skull indicate that death was from a massive blow to the top of the head while the victim was upright. The skeleton was intermingled with stones from a massive rockfall, suggesting that the victim was killed in a ceiling collapse, perhaps during an earthquake (Ralph S. Solecki, Columbia University).

and distorted by the weight of overlying layers of earth, debris, and stone, a condition typical of fragile bone remains in archaeological digs. Some of the skeletons, however, were in positions that seemed clearly to indicate a violent death. One of the skulls, designated Shanidar I, was found broken on top of a stone, with a pile of rocks above. The nature of the breakage indicated a violent blow to the top of the skull, causing the sides of the skull to burst outward (Figure 2.4). The gruesome fact is that the empty skull of a long-dead human does not build up pressure and explode in this way when struck; the owner of this skull was most likely alive and standing upright when the rock struck him. The skull, along with the broken but still articulated bones of the rest of the torso, was found mingled with a pile of fallen stone.

Professor Solecki (1971) described the find:

> As we exposed the skeleton there was increasing confirmation that this individual had been killed on the spot by a rockfall. . . . My reconstruction of this fatal accident is that the individual had been killed by a rockfall while he was standing. . . . His body was not completely covered with stones, although the impact was forceful. . . . A number of stones must have fallen on him within split seconds, throwing his body backwards full length down the slight slope.

Much discussion has focused on the health of two of the earthquake victims Solecki uncovered in the cave. Laboratory examination of the remains of Shanidar I showed that his right arm was deformed, probably from birth. Another victim, Shanidar III, showed evidence of a half-healed puncture wound to the chest. Many of the individuals were in late middle age when they died, and an analysis of the age distribution of the cave population implies that old and infirm people were more numerous in the Shanidar society than is evident for other Neanderthal sites. Even when compared to data for anatomically modern stone-age societies, the average age at Shanidar is unusually old. Solecki drew the seemingly reasonable conclusion from his findings that this society was surprisingly humane, taking care of its infirm members and venerating its elders.

Since earthquakes killed at least some of the individuals, however, we must view this age analysis with caution. These individuals were not killed elsewhere in some active pursuit and then brought back to be buried. Perhaps their infirmities kept them homebound, so that they alone were in the cave when the earth shook; maybe infirmity made them slow to react, unable to escape the cave with others before the ceiling came down on them, making it more likely that we would find the remains of the least healthy members of the society in this setting. In any case, this cave is not a simple cemetery, where the people buried constitute a representative cross-section of their group. This is one example of the importance of an adequate understanding of earthquakes and their effects to the interpretation of archaeological evidence; had we not known that an earthquake killed these people, our entire view of their society might be skewed.

UNDERSTANDING FAULTS

We cannot know whether the Neanderthals in Shanidar Cave made any direct connection between the fault scarp outside their cave and the collapse of their ceiling in earthquakes. They may have noticed that the ground along the scarp was freshly broken after an earthquake, as humans have noticed repeatedly throughout history. Although the association of earthquakes with fissures and ground disturbances was well known throughout antiquity, not until the end of the nineteenth century did people begin to understand that some of these linear features, known as faults, did not just result from earthquakes but directly caused them.

In fact, the overwhelming majority of earthquakes, known as *tectonic earthquakes*, are caused by motion on faults. Volcanic eruptions and the collapse of underground caverns can cause earthquakes, but these are generally small and cause damage only in very small regions, if at all. Since tectonic earthquakes are so much more common than any other type, in this book the word "earthquake" refers to tectonic earthquake, unless otherwise specified.

The first explicit acknowledgment of this causal relationship between faults and earthquakes dates to the Mino-Owari earthquake of 1891, in central Japan. The Japanese geologist Bunjiro Koto noticed afterward that the ground surface had ruptured, creating a scarp 80 kilometers long, which he photographed. The ground surface on one side of the line was thrown upward relative to the other by as much as 18 feet in some places. Such displacements had been seen before in large earthquakes in other places, but Koto may have been the first to make a startling proposal:

> The sudden elevations, depressions, or lateral shiftings of large tracts of country which take place at the time of destructive earthquakes are usually considered as the effects rather than the cause of subterranean commotion; but in my opinion it can be confidently asserted that the sudden formation of the "great fault of Neo" was the actual cause of the great earthquake. (Koto 1893)

It was the beginning of a fertile time in earthquake science. Three years later, Professor Rebeur Paschwitz, using a pendulum, was the first to detect ground motion (in Potsdam, Germany) from an earthquake that had occurred on the other side of the globe (in Japan), an achievement that eventually led to the creation of seismographs, and the ability to locate earthquake epicenters.

Archaeologists today are aware that faults generate most earthquakes. Understanding precisely how this happens, however, requires highly specialized knowledge, and, as in any other field, partial knowledge may lead to misconceptions. For example, if faults cause earthquakes, would it not be the case that an ancient building destroyed by an earthquake would show evidence of the fault rupture itself? We find this fallacy in the following quote from the historian Robert Drews (1993, 38), discussing possible causes for the end of the Bronze Age:

> Nor is there any archaeological evidence that any of our six sites was at the epicenter of a catastrophic quake. At each site buildings clearly collapsed, upper stories falling to floor level, or single-story houses subsiding into the streets. At none of the sites, however, was any displacement of surface levels reported. Vertical displacement, one would suppose, should have resulted in a stratigraphic step, with all the prequake strata (at Troy, for example, from VI down to I) shifted here and there to a point measurably above or below their normal altitude.

Drews's comment shows both a misconception about the significance of the word "epicenter" and an oversimplified view of how earthquake damage occurs. What Drews and many of his colleagues do not know is that, although faults do generate earthquakes, those faults extend many tens of kilometers down into the earth, and their effects extend far from their traces at the surface. This is not surprising, since few archaeology-degree programs require their students to study seismology. Therefore, any discussion of earthquakes, to make sense, must start with some background.

Plate Tectonics

Though earthquakes are unpredictable, they are not strictly random; they shape and are shaped by the structure of the earth as a whole. They are intimately related to the surface topography and deep structure of the earth through a process called *plate tectonics*, and they cannot be understood in isolation. We can, however, begin with earthquakes themselves and explore how they have helped us understand the larger picture.

An earthquake begins at a single point within the earth, where the two sides of a fault start to slip past each other, a location called the earthquake's *focus*. The focus of even a shallow earthquake is several kilometers below the earth's surface. As the earthquake progresses, more and more of the fault breaks loose and begins to slip. In large earthquakes, the area of the fault that finally slips may be huge, extending hundreds of kilometers laterally and tens of kilometers into the earth. The part of the fault that slips may or may not extend as far upward as the earth's surface. In other words, in small earthquakes, and even in some damaging ones, there may be no obvious evidence at the surface that one side of the fault slipped past the other. Finding the surface rupture of a damaging earthquake can be a nontrivial problem, even immediately after the fact.

However, we do not have to find the trace of the fault at the surface to locate the focus of an earthquake. Using modern seismometers, the shaking caused by large and even moderate earthquakes can be measured nearly anywhere, even when the energy, in the form of seismic waves, has passed through the earth's interior from the other side of the world. Seismic waves come in many types. The fastest-traveling waves, called P-waves (from *primae*, meaning "first," because they are the first waves to arrive at seismographs), are nothing more than very energetic sound waves. The second waves to arrive (S-waves, for *secundae*) transmit shearing forces through the crust. An assortment of other, slower waves travel along the surface of the earth, shaking it in various orientations, although they usually do not permanently deform the earth's surface.

The different arrival times and characteristics of these waves are what allow seismologists to determine where an earthquake originated and how the originating fault moved during the event. By measuring the lag time between the arrival of the P-waves and the arrival of the S-waves at a seismograph, scientists can calculate the distance from the seismograph to the earthquake's focus. This procedure is roughly analogous to determining the distance to a lightning strike by counting the number of seconds between the flash of the fast-traveling light and the rumble of the much slower sound waves. Because the time lag only tells us the distance to the earthquake, and not the direction, the earthquake must be recorded at three different stations before the focus can be located by triangulation.

On a map of major modern earthquake locations worldwide, shown in Figure 2.5, each dot represents the spot on the earth's surface directly above the focus of the earthquake. Remember, earthquake foci are located deep within the crust, so what the map shows are earthquake *epicenters*, the projections of the foci onto the earth's surface. Most of the earthquakes are concentrated in narrow bands that surround broad, nearly earthquake-free regions.

The bands of concentrated earthquake activity correspond closely to the physical topography of the earth. Many of the most active bands coincide with the world's most prominent mountain ranges, including the Andes and the Himalayas. Another band marks the location of the mid-Atlantic ridge, the long, continuous, sub-oceanic mountain range that bisects the Atlantic Ocean. Other mid-ocean ridges show similar earthquake concentrations. In places where no major topographical features align with the earthquake bands, there are more subtle lineations, like the San Andreas Fault on the West Coast of the United States.

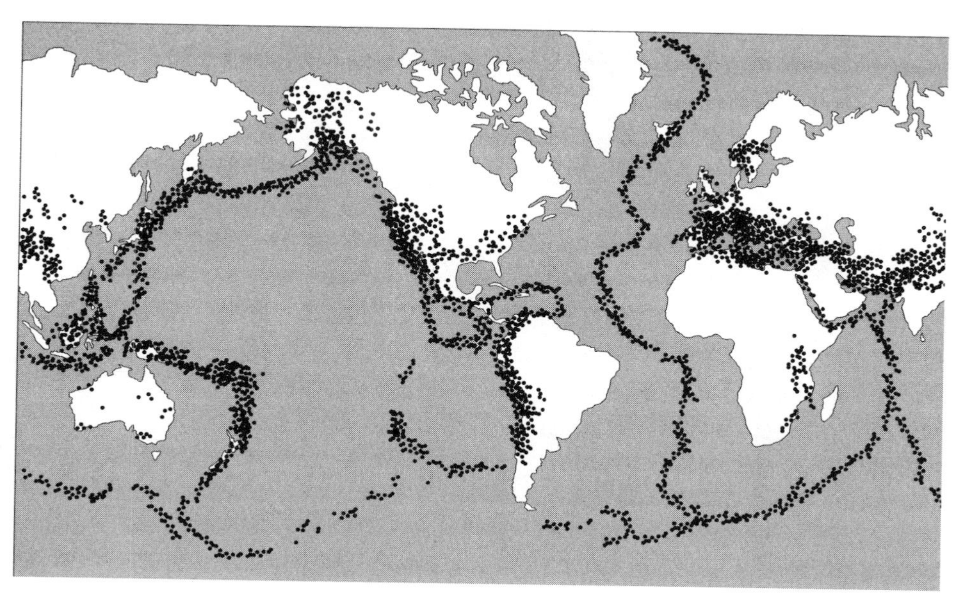

Figure 2.5. Epicenters of significant earthquakes recorded worldwide between 1900 and 1999.

Once scientists learned to use seismic waves to locate earthquake epicenters, it became clear that earthquakes were not randomly distributed but instead were linked somehow to the earth's structure. But seismic waves have also revealed less obvious information about the earth. When seismic waves pass through the earth, they encounter many different layers with different physical properties. Just as rays of light are bent or reflected when passing from air to glass, so, too, are seismic waves bent and reflected by structures with differing properties in the earth's interior, and scientists have been able to analyze those distortions to map the earth's layers.

While the outermost layer of the earth, the *lithosphere*, is relatively rigid, a large part of the earth's interior is soft enough to flow like a fluid over very long periods. Heat from the earth's interior causes slow convection currents in this softer layer. The lithosphere floats on it, moving in response to, or as part of, these ponderous currents.

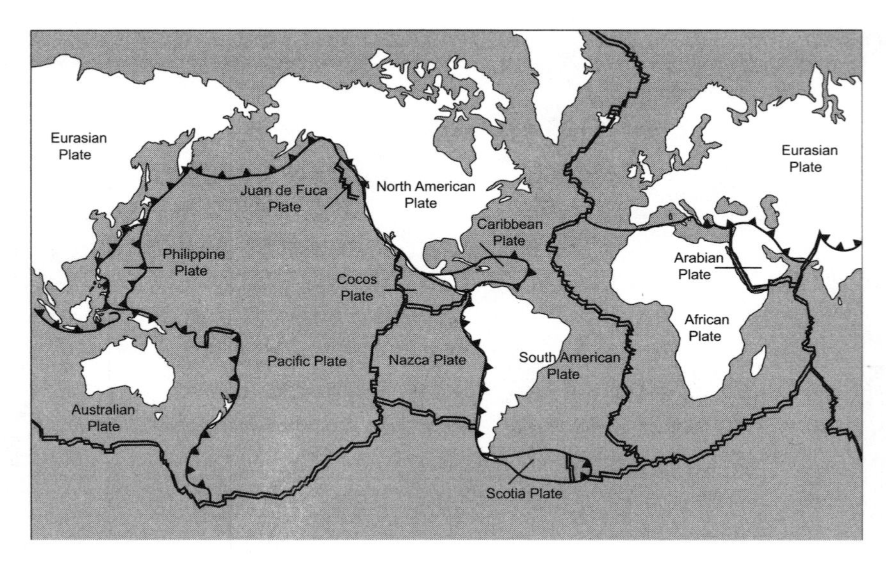

Figure 2.6 Major tectonic plates that comprise the earth's surface layer, or lithosphere. Double lines represent divergent boundaries, single lines represent transform boundaries, and "toothed" lines are convergent boundaries. At convergent boundaries, the triangles, or teeth, point toward the overriding plate; the other plate at the boundary plunges into the earth's interior.

Large areas of the lithosphere, called *tectonic plates*, are relatively quiet and stable, but they grind against one another with slow violence at their edges. These edges are where we find the largest and most continuous faults, and therefore most of the world's earthquakes. Notice how the map of the earth's major plates, shown in Figure 2.6, mirrors the map of earthquake epicenters in the previous figure.

The major plates in Figure 2.6 are all moving with respect to one another. At *transform* boundaries, like California's San Andreas zone, plates slip horizontally past one another. At divergent boundaries, plates are pulled apart, and at convergent boundaries, one plate rams beneath another. The long-term motion is slow—measured at less than 10 centimeters of offset in a year for most plate boundaries—but plates do not move smoothly past one another. After all, plate boundaries are made up of faults, enormous rock surfaces hundreds of kilometers long and tens of kilometers

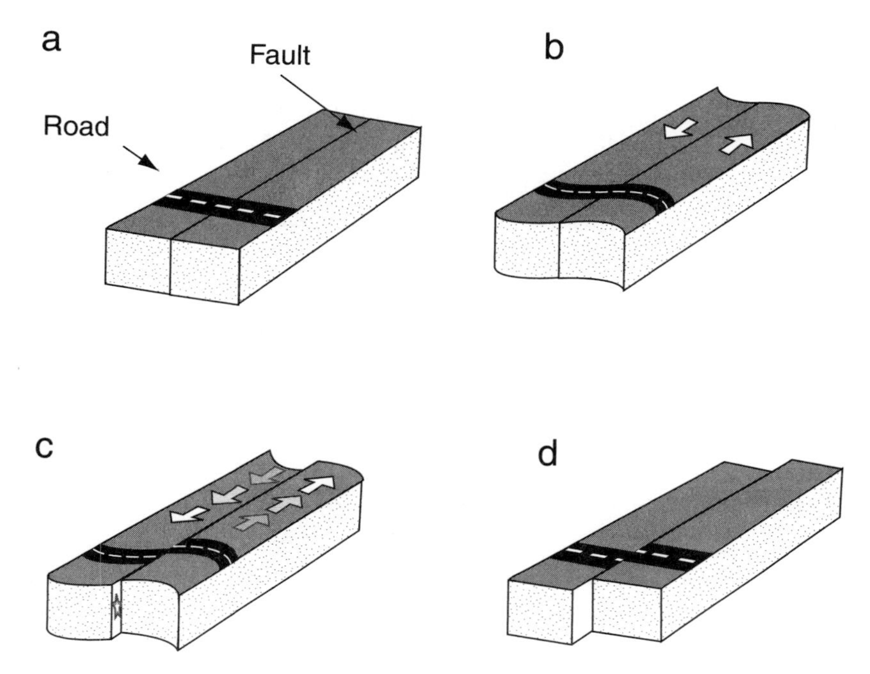

Figure 2.7 A diagram illustrating elastic rebound theory. Part (a) shows the initial, undeformed state of the crust. In part (b), the motion of the plates has caused warping of the ground (highly exaggerated for emphasis), but the fault is still stuck. In part (c), an earthquake is occurring, and the fault is slipping, releasing the energy stored by the warping. Part (d) shows the new, relaxed state, where the fault has slipped, and the road is again straight but offset across the fault.

deep, pressed tightly together by the weight of the rocks themselves. Not surprisingly, the frictional forces keeping these rocks from sliding over one another are huge. It takes a great deal of accumulated energy to make the faults move.

As plates move and boundaries are stuck, the plates themselves warp, building up internal stresses. Only when the stresses exceed the friction holding both sides of the fault together does the plate boundary slip, releasing the energy stored in the strained rocks, like a stretched rubber band releasing energy when it snaps (Figure 2.7). The energy released by the "snapping back" of the ground on either side of the fault causes the typical shaking during an earthquake.

This explanation of how earthquakes originate is the *elastic rebound* theory. In 1906, fifteen years after Koto's report from Japan, observations of the ground rupture from the great San Francisco earthquake led Harry Fielding Reid to develop this theory, which for the first time described an energy source for the tremendous destruction associated with earthquakes. Although Reid's hypothesis fell short of proposing the existence of huge, mobile plates, his description of the accumulation and release of stress in rocks "today remains a generally accepted description of the cause-and-effect relationship between faulting and earthquakes" (Brumbaugh 1999).

Fault Slip and Earthquake Size

Once part of a fault slips in an earthquake, the stress on that part of the fault is greatly reduced, and so another very large earthquake would not be expected to occur soon in the same area. Because the great plates are still moving in the same directions, however, stress slowly accumulates again until it once more overcomes the friction holding the sides of the fault together, and another large earthquake occurs.

The greater the slip and the larger the area of the slipping fault surface, the greater the magnitude of the resulting earthquake. The largest quakes on earth occur where one plate is ramming beneath another and sinking into the earth's interior, at convergent boundaries called *subduction zones*. This convergence creates a wide zone of deformation at the earth's surface. The crumpling of the overlying plate creates a chain of mountains parallel to the plate boundary, mountains riddled with faults that generate many large earthquakes. The classic examples of such mountain chains are the Andes range along the west coast of South America and the nearly contiguous Cordillera of Central America, which are of similar origin. Other examples around the world are the Himalayas, the Cascades, the islands of Japan, Sumatra, Crete, and Cyprus, and the Peloponnesos of Greece. The faults associated with these collision

Earthquake Instability

The Slider-Spring Model

In every recorded earthquake, a given area of a fault plane ruptures, and the sides of that fault plane slip past each other by a certain distance. After this occurs, we can estimate these parameters—the size of the rupture area and the fault slip—fairly accurately by analyzing the earthquake waves and aftershock patterns. However, explaining the physics behind these parameters is trickier. Why does a given area of a fault slip rather than a greater or lesser area? What determines how far the fault will slip before the motion suddenly stops? Since the area and slip of an earthquake determine its magnitude, predicting these quantities for a given fault would be of great use in estimating seismic hazard.

One model used to explore this question is the *slider-spring instability model*. Like a heavy block on a tabletop, the two sides of a fault can withstand significant stress before they slide past each other; friction locks the two surfaces together until the stress exceeds the frictional resistance. In the figure shown here, we have attached a spring to a heavy block and then stretched it until the block slips; the amount the spring stretches is proportional to the force applied to it. The complicating factor is that the force required to keep the block sliding is less than the force required to start it moving in the first place. This is because the friction between two sliding surfaces (*dynamic friction*) is lower than the friction between two surfaces at rest (*static friction*), so that, once the block slides, it continues to slide for some distance until the spring contracts. When the tension in the spring is no longer sufficient to keep the block sliding, the block stops; once this happens, the static friction again must be overcome before the block can be set in motion once more.

This start-and-stop motion bears great similarity to the episodic motion of faults during earthquakes, and the equations governing the slider-spring system have been modified to study the unstable

continued on page 48

behavior of earthquake-generating faults. This, by the way, is also the principle behind the uncontrollable motion of a skidding car; once the tires begin to slide, their grip is governed by the weaker dynamic friction.

A greatly simplified version of a fault model is illustrated in the figure; a real fault would be modeled as a large collection of linked blocks and springs. In panel (a), the spring is stretched by its post-slip amount, with force F1, right after a previous slip episode. In (b) the spring has been stretched enough to overcome static frictional resistance (F2). In (c) the spring is back to the immediate post-slip stretch.

The stretching of an ideal spring is governed by the equation $F = kx$, known as Hooke's law, where F = force, k = the spring constant (reflecting the stiffness of the spring), and x = the distance the spring is stretched. In the figure, the block's total displacement is the difference between F2 and F1, divided by the spring constant, as shown in the following equation:

$$x = \frac{F2 - F1}{k}.$$

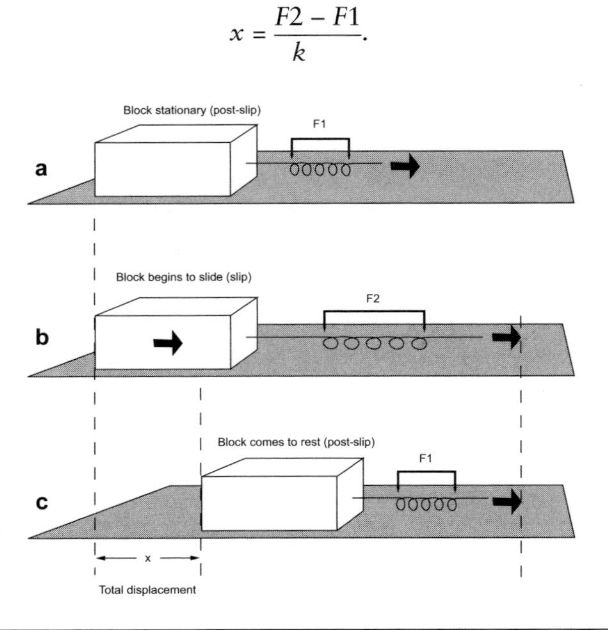

zones are *reverse faults,* or *thrust faults*; they slant, like ramps, and rocks on one side of the fault are pushed up and over the other side (Figure 2.8a).

The earthquakes that comprise the next largest category are those that occur on *transform*, or *strike-slip*, faults, boundaries where two plates slide horizontally past each other (Figure 2.8b). Although very little topographical deformation may take place along a pure transform fault, the enormous forces driving the plates create wide *shear zones*, where rocks are smeared and fractured as they are ground between the two plates. Existing mountain chains, rivers, and other linear surface features that cross the shear zone become offset by its motion over time. The San Andreas Fault zone is one of the most famous of these transform boundaries; others include the Dead Sea Transform in the Middle East, the North Anatolian fault in Turkey, and several faults in Mexico and Guatemala.

The third type of plate boundary is the divergent boundary, or *rift*, where two plates are moving apart from each other. Although this kind of boundary can have a spectacular effect on topography, leaving huge, deep valleys or lakes—and eventually oceans—in the gap between the plates, the earthquakes such boundaries generate are somewhat smaller than those generated in the other two situations. This is because the earth's lithosphere is thinner and weaker in rifts than elsewhere, and stresses are smaller. The faults associated with these boundaries are called *normal faults*. Like reverse faults, these faults are inclined like ramps, but in this case, one side slides *down* the ramp formed by the other (Figure 2.8c). A classic example is the Great Rift Valley in East Africa. Other examples include the Red Sea and the Imperial Valley in California, which is the landward extension of the Gulf of California in Mexico.

Plate boundaries, however, are seldom purely one type or another. Two plates may be sliding past each other while at the same time separating slowly, as happens on the Dead Sea Transform; this separation leads to subsidence, which explains why the Dead Sea has the lowest elevation on earth, at 411 meters below sea level. In other places one plate may slide beneath another at an angle, so that both reverse and transform motions occur on the faults.

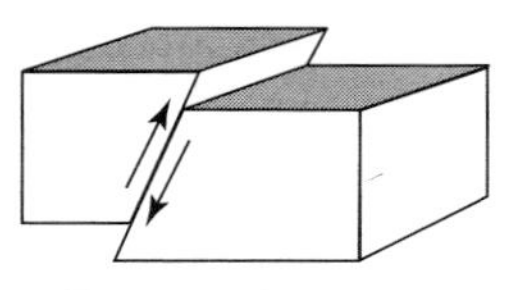

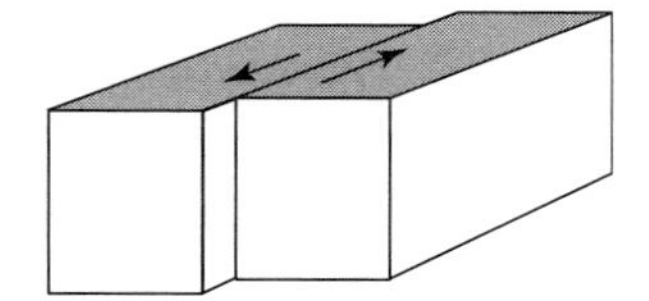

b. Transform fault

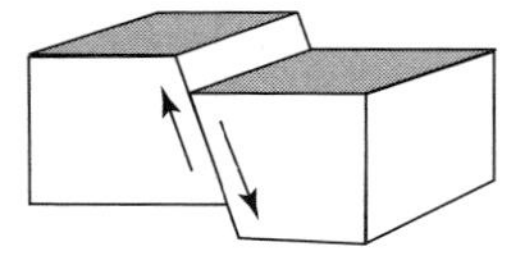

c. Normal Fault

Figure 2.8 Types of faults: (a) *reverse faults* (or *thrust faults*) are inclined planes of weakness in the earth's crust, where the rocks on one side are thrust upward over the rocks on the other side, as if being pushed up a ramp; (b) in *transform faults* (*or strike-slip faults*), one side of the fault slides horizontally past the other; and (c) *normal faults* are inclined like reverse faults, but the rocks on one side slide down the "ramp" relative to the rocks on the other side.

Furthermore, most faults are not perfectly straight, and while motion on one part of the fault may be purely strike-slip, a slight bend in the fault can cause separation or convergence on other parts, resulting in small, localized pull-apart basins or so-called push-up mountains (Figure 2.9). California is full of such features, where lakes or mountains form along bends in the San Andreas Fault.

Further complicating the matter, often plate boundaries are made up not of single faults but of entire zones of related faults that cross one another, divide, and coalesce. The motion on one fault of the system may be very different that that on a neighboring fault, although both faults are part of the same plate boundary.

This variability in fault mechanism is responsible for the varied terrain in many regions. For example, Israel's Carmel-Gilboa mountain range is located very near the Dead Sea Valley, the

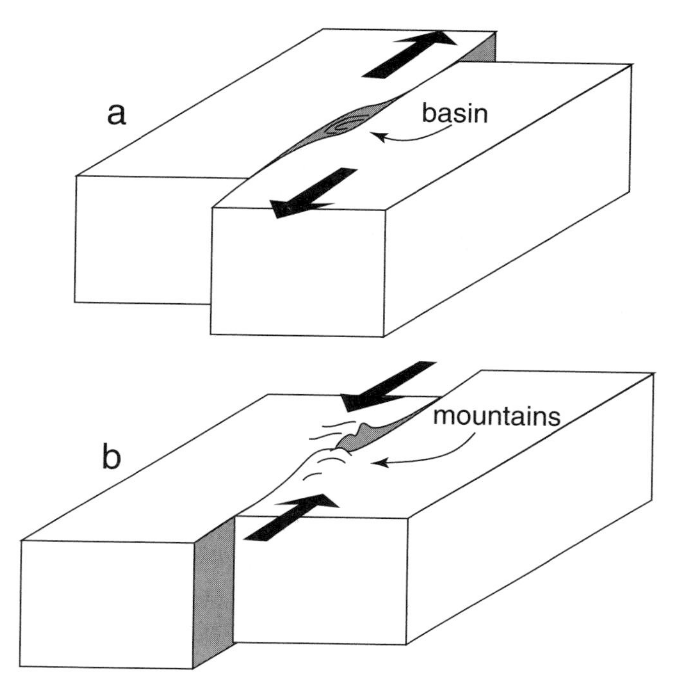

Figure 2.9 This diagram shows the effects of motion on a transform fault that is not straight. Depending on the geometry and fault motion, the bend in the fault can be either in tension (a), creating a *pull-apart basin*, or in compression (b), creating *push-up mountains*. Examples of both these features can be found throughout California along the San Andreas Fault.

deepest valley on earth. The valley was formed because the Dead Sea Transform has a component of pull-apart motion, creating a slight rift where the two plates part company. However, secondary fault systems branch off from it, and the motion on some of these faults is reverse; this reverse motion has created the Carmel-Gilboa mountain range. This complicated interaction of fault zones exists worldwide and can make the geophysical interpretation of geography very difficult.

Tectonic shaping of the land has its effects on human constructions as well. The oldest example I know of is in Ubediyeh, Israel (Bar-Yosef 1993), in a little gully south of the Sea of Galilee, near

Figure 2.10 The 1.5-million-year-old pebble floors of Ubediyeh, Israel. The floors were horizontal when they were created, probably by members of Homo erectus pressing pebbles and animal bones into the mud. The floors have been tilted approximately 60 degrees from the horizontal by thousands of earthquakes (Courtesy Institute of Archaeology, Hebrew University).

the Jordan River (Figure 2.10). The "structure" is the simplest one possible—a floor made of pebbles—put there by the appropriately named "Pebble Man," thought to have been a member of the species Homo Erectus. This is, in fact, the oldest known hominid construction on earth, made some 1.5 million years ago. Apparently, Pebble Man made this floor by pressing river pebbles, parts of stone artifacts, and even bones of now extinct mammals into what was a muddy bank of the Jordan River. Thus, instead of working and sleeping in the mud, Pebble Man created a relatively stable, dry base for his operations.

When this site was first excavated, the nature of the surface was not immediately clear, since it is inclined 60 degrees from the horizontal. In a rare collaboration, geologists and archaeologists eventually determined that the surface had been horizontal, or nearly so, when built. The floor's present tilt is the accumulation of earthquake tilting during the 1.5 million years since the floor was created. Because modern data suggest that large earthquakes recur an average of every four hundred years in this area, some three thousand to four thousand such events have occurred, each tilting the floor by the miniscule average of just two hundredths of a degree.

The frequency with which large earthquakes will strike a particular region depends on the amount of motion on a plate boundary, the type of motion (normal, reverse, or strike-slip), the sizes of the faults in the area, and the interactions between faults. The *recurrence interval*, that is, the time that passes between quakes of a given size, is critical to estimating the seismic risk to a region. In areas where large quakes are likely, but where the recurrence interval is very long (more than one or two generations), it is very difficult to maintain earthquake awareness in the populace, and very difficult to establish and enforce seismic building codes. When people see that ancient local buildings have never been destroyed by earthquakes, they find it easy to dismiss the danger, even if seismologists warn that there are active faults in the region.

A powerful example of this potentially tragic situation occurred in the city of Bam, in Iran. This city was home to the two-thousand-year-old citadel of Arge-Bam, which was destroyed by an earthquake on December 26, 2003, along with 90 percent of the buildings in the modern city. More than twenty-six thousand people were killed, and thousands more were injured, in the earthquake. Although the ancient buildings indicated a very long recurrence interval between damaging earthquakes in Bam, the 2003 destruction revealed evidence that, after an ancient quake, some of the citadel walls had been repaired (Earthquake Engineering Research Institute 2004). Scientists are still trying to estimate the recurrence interval in that area, now that this tragic event has highlighted the seismic danger.

The focus of the Bam earthquake was almost directly under the city, at the relatively shallow depth of 7 km. Thus, even though the earthquake was only moderately large, the damage was severe but confined mostly to the city of Bam. The damage would have been more widespread had the earthquake been larger, with a deeper focus.

In an archaeological site, determining the size of an ancient earthquake requires knowledge not only of the damage at the site but also of the types of faults in the area and the regional nature of the damage. In the Shanidar Cave example, none of these other pieces of information was available. The fault Solecki noticed did not run directly through the cave, and there was no evidence of cumulative ground displacement or tilting within the excavation, like in Ubediyeh, but Solecki attributed the collapse of the cave ceiling to an earthquake on this fault.

In reality, the epicenters of the Shanidar earthquakes could have been farther away. There are many faults in this region, and a large earthquake on a more distant fault may cause more damage than a smaller quake on a nearby one. For example, the greatest damage from the Loma Prieta earthquake described earlier was found not near the epicenter in the Santa Cruz Mountains but nearly 100 kilometers away, in the Marina District of San Francisco. This is a reminder that most of the damage in earthquakes occurs not because of faulting or permanent ground displacement but because of the transient shaking caused by large seismic waves radiating outward from the focus as the fault slips.

That is not to say that archaeologists should ignore ground displacement, which remains one of the few incontrovertible signs of earthquake damage. They should be aware, however, that the chance of observing such displacement in an archeological site, even one destroyed by an earthquake, is very small. It is a remarkable archaeological and geological discovery when we unearth an ancient wall that has been bisected by fault motion; however, it is usually more rewarding to search for the more common types of damage caused by shaking.

What kind of damage should we expect from shaking? What is the biggest earthquake we can expect in a given region? How large an area might a single earthquake affect? How often do earthquakes recur? These are the kinds of questions that must be answered before informed decisions can be made to either assign damage to earthquakes, or dismiss earthquakes as a hypothesis. Fortunately, scientists and engineers now have a fairly good understanding of the sorts of damage to expect in earthquakes and how damage areas relate to earthquake size. For areas with good historical records of seismic activity, and where faults are well mapped, we also have reasonable estimates of earthquake risk, including the maximum probable size of an earthquake in the region and the likelihood of large earthquakes in the future. However, not all areas enjoy such good historical records, and even the best records do not always extend far enough into the past to give a complete picture of regional seismic hazards. This is where archaeology could potentially provide a valuable service; by extending a region's earthquake record into prehistory, we could have a more reliable estimate of how likely large earthquakes would be in the future.

MAGNITUDE AND INTENSITY

Earthquake size can be understood in several different ways. The most measurable and straightforward is earthquake *magnitude*, which is traditionally a measure of the maximum amplitude of shaking in an earthquake, a measure that can be correlated to the amount of energy released in the quake. Many scales are used to measure magnitude, the most famous being the Richter Scale developed by Charles Richter in 1935. This is the most familiar scale to the general public, but it is not particularly useful for very large earthquakes. It tends to saturate at high magnitudes so that all large earthquakes look rather alike, even if the energy released is very different. A more useful indicator is the *moment magnitude*, calculated from actual fault motion, which gives approximately the same numbers as the Richter magnitude for small earthquakes but more successfully distinguishes between big ones. When it is available or can be estimated from historical data, the moment magnitude is the number used in this book.

All the magnitude types are open-ended measurements, that is, they can be arbitrarily small or large. Practically, however, the magnitude is limited by the amount of energy that can be released, which in turn is limited by the size of the fault that can rupture and

Earthquake Magnitude

Calculating the magnitude of an earthquake is a complicated problem. The first scale developed for doing so was the local magnitude, M_L, published by Charles Richter in 1935. Richter defined the local magnitude by measuring the maximum amplitude (measured in thousandths of a millimeter) of waves recorded by a Wood-Anderson seismometer situated 100 km from the earthquake's epicenter, and then taking the logarithm of that amplitude. The Wood-Anderson seismometer was the state of the art in 1935 but is no longer used, so other types of seismograms are converted to determine the local magnitude.

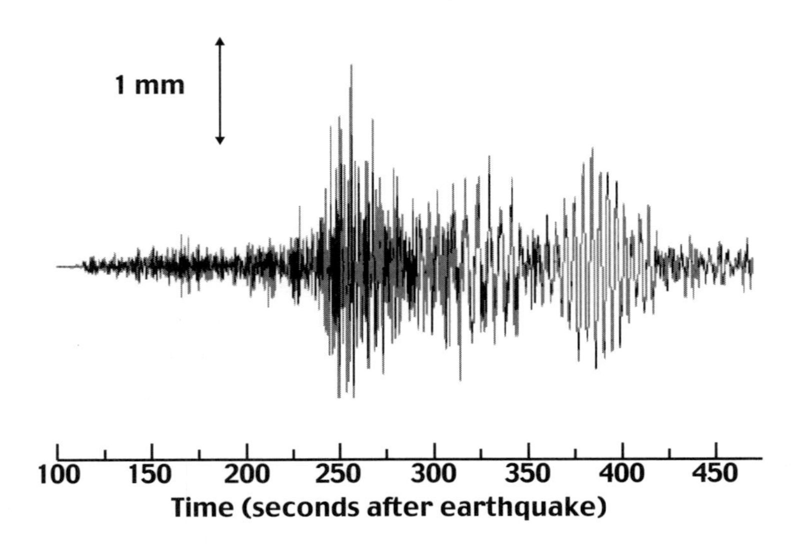

Since it uses the logarithm of the amplitude, the magnitude scale can describe a huge range of earthquake magnitudes without using extremely large or small numbers. The shaking in a magnitude 6 earthquake is ten times greater than in a magnitude 5, and a hundred times greater than in a magnitude 4. The energy difference is even more striking: for every unit increase in magnitude, the energy released in the earthquake increases by a factor of 32.

However, the local magnitude is not always the best indicator of an earthquake's "size." For example, the local magnitude is not very sensitive to differences between extremely large earthquakes, above magnitude 7. Moreover, the depth of an earthquake greatly influences the amplitude of shaking at the surface, so that very deep earthquakes have smaller local magnitudes than shallower ones that release the same energy. Therefore, other magnitude scales have been developed to answer specific questions about earthquakes. For instance, scales can describe amplitudes based on P-waves (body-wave amplitude, or m_b), on surface waves (M_s), or on seismic moment (M_w), which I describe below.

The seismic moment magnitude was developed by scientists interested in describing the characteristics of earthquakes' sources uniformly, for a wide range of source depths and earthquake sizes. The calculation of the moment magnitude is not as straightforward as measuring the amplitude of waves on a seismograph. Instead, scientists use either field measurements or wave analysis to determine the size of the fault plane rupture, the strength of the rocks, and the amount of displacement between the two sides of the rupture surface. Scientists then use these values to calculate the seismic moment, which provides a good description of the total size of an earthquake, from the very large to the very small. Since most nonscientists are unfamiliar with the concept of moment, the seismic moment is converted into a logarithmic magnitude scale expressed much like Richter's local magnitude.

the amount of strain energy the rocks can store before they break. Thus, at least on this planet, we do not expect to see earthquakes with moment magnitudes larger than about 9.5.

In any given region, small earthquakes are much more common than large ones. In fact, we in geophysics have discerned a simple statistical rule that seems to govern the occurrence of small and large earthquakes, called the *frequency-magnitude relation*. The equation for this relation is

$$\log(N) = A - BM$$

where N is the cumulative number of events, B is a constant that depends on the overall seismic risk in the region, and A is a constant related to the size of the area sampled and the length of time for which the data have been collected. Simply stated, the equation implies that if geophysicists accumulate a record of small earthquakes in a region, they can estimate how often, *on average*, larger earthquakes will strike. The average time that passes between earthquakes of a given magnitude is called the *recurrence interval*; however, it is important to recognize that this is a statistical measure, and cannot predict the actual date when an earthquake will strike. Furthermore, we must estimate independently the maximum expected magnitude using the lengths of known faults in the area, which can be difficult if the faults have not displaced the earth's surface relatively recently.

Earthquake magnitude is the number cited by news agencies and in most lists of recent earthquakes, but it is of less direct interest for archaeologists (and for people living in earthquake hazard areas) than the *intensity* of an earthquake. *Intensity* is a measure of the amount of damage the shaking causes to buildings and landforms, and the extent to which the earthquake disrupts human activity. Obviously, this scale depends not only on the earthquake but also on population density, building construction methods, and ground conditions in the affected area. The most commonly applied intensity scale in the United States, and the one used in this book, is the Modified Mercalli Intensity (MMI) scale, which was developed to reflect conditions in California, the area of the U.S. where people are most strongly affected by earthquakes. (Alaska has more seismic activity, but its small population leads to less publicity and less research attention.) The MMI scale ranges from I to XII, where I is rarely felt, except under special circumstances, and XII signifies total destruction. The scale is somewhat subjective; the following description applies to intensity VII effects: "Everybody runs outdoors. Damage negligible in buildings of good design and construction; slight to moderate in well-built ordinary structures; considerable in poorly built or badly designed structures; some chimneys broken. Noticed by persons driving cars" (Wood and Neumann, 1931). This description is similar to the one for intensity VIII on another scale commonly used in Europe, the Rossi-Forel scale, which has only ten divisions.

Obviously, the extent of the described damage depends greatly on the definition of "well-built ordinary structures." The ordinary structure in modern California is quite different from the ordinary structure in prehistoric Turkey, and so we should expect much greater total damage in ancient Troy than in Los Angeles, given the same amount of shaking. Even today, construction methods differ greatly in various earthquake-prone countries, and moderate earthquakes often cause much greater damage in the Middle East or Eastern Europe than in California or Japan, where earthquake building codes are quite stringent.

Intensity is intimately connected with magnitude, of course. For example, very small earthquakes, such as the magnitude 3 earthquakes that occur daily on the San Andreas Fault system, go undetected by the general populace. For this magnitude, we never expect to see intensity VII areas, even if buildings were erected on foundations made of Jell-O right at the epicenter of the quake. For larger earthquakes, however, we may or may not see such high intensities, depending on the kinds of soil prevalent in the area. The depth of the earthquake focus also plays a large role; deeper earthquakes have lower intensities than shallow ones and spread out over larger regions. In general, though, severe damage is possible in the epicentral area of an earthquake of magnitude 6 or higher. Larger earthquakes usually have wider areas of severe damage, with more severe and total damage near the epicenter.

Any city on a significant fault zone will have been rattled many times in its history. If that history is long enough, the city will certainly have suffered significant damage from earthquakes both directly beneath it and further away on the fault. Even cities far from an active fault can experience severe damage, if conditions are right.

Measurements have shown that certain topographical features, such as narrow ridges and small hills, can resonate with earthquake waves, causing more violent shaking and more severe damage than would generally be expected. This effect has been studied in detail for a hill in Tarzana, California. Measurements taken there during aftershocks of the January 17, 1994, Northridge earthquake show that ground motion at the top of the hill was as much as 4.5 times greater than at the base. Part of this resonance is thought to be the result of the internal structure of the hill, since theoretical calculations based on its shape alone predict only a twofold increase. Nonetheless, even doubling the shaking is a significant effect.

Even where topography is flat, areas composed of poorly consolidated materials, like loose, water-deposited sediments or improperly compacted artificial fill, suffer disproportionately severe damage in moderate earthquakes. In the 1989 Loma Prieta earthquake, homes built on artificial fill in San Francisco's affluent Marina District were badly damaged when the unstable soil liquefied during the shaking. The earthquake's epicenter was about 90 kilometers away, but damage in the Marina District was worse than in most places closer to the epicenter.

In the Mediterranean area and the Middle East, the Mercalli Intensity Scale has particular ramifications for archaeology, because archaeological sites in arid regions generally combine the worst of both worlds. When settlements must be located near a reliable source of water year-round, say, near a well or a spring, these sites will have been occupied repeatedly over the millennia. As a result, the most important archaeological sites are usually marked by tells (*tel*, in Hebrew)—mounds made of the accumulated debris of successive civilizations (Figure 2.11). With some rising tens of meters above the surrounding landscape, tells were formed when, following the destruction of a city, the succeeding residents salvaged what they could from the rubble, spread out the rest, and built on the resulting heap. Thus, the buildings on top of the pile suffer both the potential for resonant amplification, because of the hill's shape, and the shaky footing of foundations built on uncompacted fill.

In addition, many of these sites were founded originally on ground very susceptible to seismic shaking. In desert regions with mountainous topography, most of the precipitation falls in the

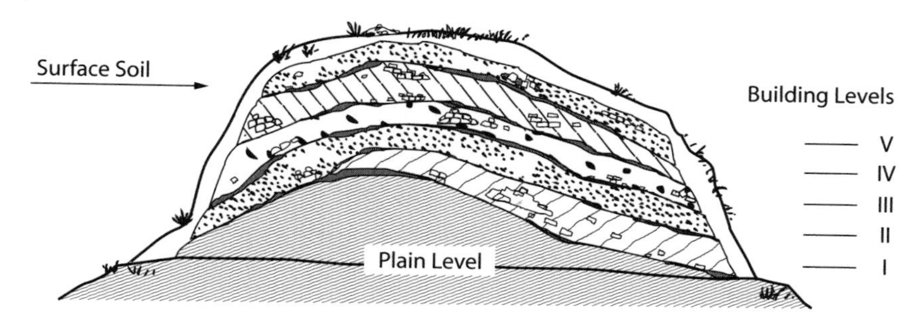

Figure 2.11 A diagram of the structure of a typical tell.

mountains and collects in streams that flow down to the lower areas, where they deposit sediments eroded from the higher elevations. Over time, this creates great wedges of loose sediment, called *alluvial fans*, which in many ways are ideal sites for settlements. The sediments are loose and easily worked, slopes are gentle, and fresh-water springs are common where water percolates down from the mountains. The mountains themselves, however, are the result of activity on faults that often are still active. Indirectly and directly, then, earthquakes entice people to build in the region and repeatedly destroy their handiwork.

To escape destruction in such inauspicious conditions, a structure must be very carefully designed, with great strength and flexibility relative to its mass. Unfortunately, throughout the Middle East and the Mediterranean region, this has not been the case until only quite recently. The traditional building method for grand public buildings has been unreinforced masonry, which has very little strength under the kinds of horizontal stresses that earthquakes cause. The inertia of the stone blocks prevents these massive structures from moving in concert with the shaking ground, and their rigidity means they crumble rather than flex under the stress that results. Their massiveness ensures that, when they do collapse, anyone inside is almost certain to be crushed.

Humbler structures in these regions are generally made of mud brick. Since mud brick is not very strong, builders compensate by making walls very thick and, consequently, very heavy. Though popular, partly because buildings stay cool during hot daytime temperatures, mud-brick construction is even more likely than stone to collapse in an earthquake. These construction practices are the reason even relatively small earthquakes like the one in Bam often cause such misery in the Middle East today. Since these are ancient practices, we can be sure that the same has been true in the past.

JERICHO, 1927

Jericho is one place where the sad combination of poor construction and seismic hazard has persisted throughout history. An attentive observer can easily follow major faults along the Jordan River plain (Reches et al. 1981). They disrupt the flat sediments in many places, and create lines of springs where groundwater flows up along the fractured fault plane to the surface. During the 1927 Jericho earthquake (discussed in more detail later), the ground actually cracked in several places, and water poured out. The shaking during an earthquake can change water-saturated sand or silt into a fluid that flows up to the surface during an earthquake, creating mud or sand "volcanoes." In other cases, the water squeezes out of disturbed rock or sand because of changes in pressure and in the permeability of the ground, forming new springs.

Jericho was severely damaged in the 1927 earthquake, which was centered in the northern Dead Sea Basin on the Dead Sea Transform. Jenin, Nablus, and other towns in the region also suffered greatly. Some 1,000 people were killed, a significant number in this region of Palestine, which was only sparsely populated at that time. Only around 250,000 people lived there then, versus 9 to 10 million today.

A great deal of this damage can be attributed to the poor construction practices that were the norm in the region, with many cities built on a base of alluvium (loose, water-deposited sediments), with several layers of previous quake remains as a foundation. The geologist Bailey Willis (1928) described the effects in Nablus:

Figure 2.12 A panoramic view of Tel Jericho as it looked when Kathleen Kenyon was leading excavations there. The ancient site is the long, low mound in the center of the photo, with modern houses in the foreground and background. The contrast between the oasis of Jericho and the surrounding desert makes it obvious why this site has been inhabited off and on for nine thousand years (Photo Archive Submitter/National Geographic Image Collection).

> Nablus was divided into three sections. Two of them, the northern and southern sides of the town, were scarcely damaged. The central strip was very badly ruined. The former stood on the rocky slopes on either side of the little valley in which the town is situated. The middle zone was piled up on the alluvium and on the debris of former earthquake ruins, an unstable support for massive walls of mud and stone, three or four stories high.

Although the modern city of Jericho is built on several layers of older ruins, it does not occupy the site of the ancient biblical city. The mound of Tel Jericho, the ancient site shown in Figure 2.12, is now deserted. Archaeological evidence indicates as many as twenty-two levels of destruction within this mound. It should not be surprising to find that ancient Jericho was particularly susceptible to earthquakes, probably even more so with each successive layer. In fact, many details of the biblical story of Joshua's attack on Jericho, with its famous collapsing walls, have striking parallels with damage recorded after historical earthquakes in this region.

Before I describe the similarities, though, I must address the reliability of the Bible as a historical source. Just how much we can infer from ancient written records, however, is one of the most difficult topics in earthquake archaeology, and not just for Jericho or other sites of biblical significance. Unfortunately, the human written record is not as unambiguous as a trace from a seismograph or a number on an intensity scale, and the passage of time has only made the problem worse. The challenge, then, is to take what we know about earthquake science today, and correlate it when possible with both written history and the archaeological record.

CHAPTER 3

History, Myth, and the Reliability of the Written Record

> And it came to pass, when the people heard the sound of the trumpet, and the people shouted with a great shout, that the wall fell down flat, so that the people went up into the city, every man straight before him, and they took the city.
>
> —Joshua 6:20

The story of Joshua's battle at Jericho is famous and familiar from the Hebrew Bible: the mystical march seven times around the city, the blowing of the horns, the mighty shout, and the collapse of the walls; the feat has been immortalized not only in the Bible but in popular literature, art, and folk songs. However, visiting this famous archaeological site is actually disappointing. When many years ago I first saw Jericho from a distance, it looked like nothing more than a dusty mound some 30 feet high, 1,000 feet long, and 400 feet wide, covered with dry weeds. Only in early spring is it green for a few weeks. It looked even more dismal on closer inspection. Past excavations had left behind seemingly haphazard trenches, with piles of dirt beside them. In some of the trenches—I presume ones in which the excavators decades ago had tried to protect important findings—shreds of old plastic sheeting flapped in the wind.

Modern times have not improved the site. In the mid-1990s, builders plunked a massive steel and concrete cable-car station on the southern end of the mound. The structure—with its foundation partly on top of the ruins—makes the ancient mound look even less impressive. Not far away is a modern gambling casino—a shared Palestinian and Israeli project that attracts hordes of mostly Israeli gamblers between intifada periods.

However, Jericho has always had one advantage: ever since its first occupation more than nine thousand years ago, it has boasted a dependable, year-round source of clean water, the spring of Elisha. Also known as the Sultan's spring, it has been used since antiquity to irrigate the lush, agricultural oasis that the mound overlooks. One can certainly see why this isolated island of green in the harsh desert of the Jordan Valley would have been coveted by the Israelites when they saw it, for, according to the Bible, God promised them Jericho and all the lands west of the Jordan.

Certainly, the Bible is far from a neutral history. First, the stories in the Old Testament were not committed to writing until the first millennium BC—some seven hundred to a thousand years after the events they describe are presumed to have taken place. How reliable can such an extended oral transmission be? Second, the Old Testament's aims—to glorify God and secure the cultural identity of the early Israelites—surely must have superceded concerns about historical accuracy as the stories were passed down. The same is true of the story of Troy as told in Homer's *Iliad*, and of any history long maintained in an oral tradition; the biases of generations of storytellers diffuse into and taint the record of events. In fact, in many places in the Bible, it is evident that separate events are combined, reordered, or otherwise altered, and sometimes even repeated in more than one historical context, either to suit political or religious aims or simply because years of retelling introduced distortions. Still, the biblical story of Joshua's conquest of Jericho contains many odd details that would make a great deal of sense if the destruction of Jericho's walls were caused by an earthquake.

This is not a new proposal, but it is one that invariably annoys people at both ends of the religious spectrum. Religious fundamentalists—Christian, Jewish, or Islamic—see heresy in any attempt to assign natural causes to the miracles of the Old Testament. Their beliefs depend on the assumption that events transpired exactly as described in their shared holy scriptures. On the other hand, some archaeologists are nearly as uncomfortable as fundamentalists are but for the opposite reason. They consider the Bible as nothing more than a religious text, written as support for religious and political views. Using the Bible as literal history invites instant criticism from these archaeologists, known as biblical minimalists, who dismiss the historical accuracy of the Bible altogether.

One example of the inaccuracy that has crept in through the retelling of biblical stories is apparent in the books of Joshua and Judges. Both books describe the crushing defeat and slaying of the legendary Jabin, King of Hazor, the first telling how he was slain in a glorious battle led by Joshua against the city of Hazor, and the second relating how, several hundred years later, he was again killed (and Hazor again razed) by Deborah and Barak. Clearly, the slaying of Jabin was a pivotal military and moral victory for the Israelites, and credit for the victory was parceled out over several generations.

Confronted with such inconsistency, it is not hard to believe that the sequence of events surrounding the battle of Jericho might also have been altered through centuries of retelling, especially when the purpose of perpetuating the story was, one assumes, to glorify God and sanctify Israel's claim on the region. Indeed, some biblical scholars have gone so far as to question whether Joshua actually existed, or whether his conquests were the work of several commanders, compressed and aggrandized over time. The dual slaying of Jabin would lend credence to this idea. Still, most historians and archaeologists agree that dismissing the Bible as historically valueless is extreme. Time and again archaeologists have uncovered evidence that events mentioned in the Bible have some basis in fact, although the chronology or the details may have been changed. The destruction of Jericho is a case in point.

The city of Jericho is near the Dead Sea Transform, and even nearer to one of its branches, the smaller Jericho Fault. We know that this region has been subject, historically, to frequent, devastating earthquakes, such as the one in 1927. Conceivably, then, an earthquake could have struck during the battle at Jericho. The story of Joshua's battle and the collapse of the walls *could* have happened as told in the book of Joshua, with some help from an earthquake; or, perhaps more likely, the attack on Jericho may have been opportunistic, triggered by the collapse of the city's defenses in an earthquake. Either way, had an earthquake indeed struck the area and caused the collapse of Jericho's walls, the story that ended up in the Bible might have been muddled enough to partly obscure that fact. For the earthquake scenario to be anything but pure speculation, however, we can only turn to archaeology.

Evidently, several levels of Jericho collapsed in catastrophic events, which the excavators agree were probably earthquakes. None of those layers is architecturally very impressive—certainly not remotely close to the grandeur of ruins from Egypt, Mesopotamia, Western Turkey, Greece, or Rome. Collapsed walls of mud brick and crudely cut or uncut local stones are interspersed with brown and black ash layers—the remains of fires that accompanied the periods of destruction. Whether any of those layers is associated with Joshua's conquest of Jericho, however, is debatable for many reasons.

Biblical historians have traditionally placed Joshua's attack on Jericho somewhere between 1250 and 1400 BC, arguing that the Egyptian evidence from that time best correlates with the biblical story. But the archaeological record for that interval at Jericho is plagued by controversy. There is evidence that human habitation at Jericho is longer-lived than at any other city in the world yet excavated, reaching back nine thousand years before the present. One early expedition to the ruins reported excavating walls from Joshua's time. According to Kathleen Kenyon (1979), however, the archaeologist responsible for the most complete excavations in Jericho in the 1950s, the partial remains of a single house is all that remains from the period that might relate to the biblical figure of Joshua; the rest of that level has eroded away, along with any evidence that could corroborate or refute the biblical account. However, Kenyon advanced the theory that the Exodus, and the stories associated with it, may have been a synthesis of up to three separate migrations, spanning from the time of the Hyksos in

Egypt (the end of the seventeenth and beginning of the sixteenth centuries) up to the more traditional dates mentioned above.

I discuss the archaeology of Jericho and the surrounding towns in more detail in chapter 7, after describing, in chapters 4 and 5, some of the signs that earthquakes leave behind in the rubble. For every question archaeology answers, however, it raises two new ones, and previous excavations have muddled the strata that remain for interpreting those new questions. Archaeology in Israel and Jordan has always faced this difficulty.

Early excavators, almost without exception, began their work with the goal of uncovering the remains of a biblical event or location. This inherently biased approach led to many problems, not the least of which was the dismissal and subsequent destruction of any layer the excavator considered irrelevant to the quest of the biblical target. Layers above the excavators' targets were shoveled unceremoniously into wheelbarrows and buckets, and dumped over the edges of tells. Sometimes, misled by assumptions of what the layer of interest would look like, excavators destroyed the very strata they were seeking, focusing instead on underlying layers that actually predated the events they sought. The firmness of purpose and narrow goals of these early "archaeologists" in the Holy Land destroyed many centuries of irreplaceable data.

It may be some consolation to modern biblical archaeologists that their predecessors were not alone in this destructive single-mindedness, for single-mindedness was a defining feature of the early period of archaeology, where the discovery of financial or cultural treasure was the driving force behind most excavations. Evidence of biblical events was only one type of cultural treasure. Another oral tradition sparked similar fervor, although among a different subset of investigators.

THE SEARCH FOR HOMER'S TROY

Like the Bible, Homer's great works, *The Iliad* and *The Odyssey*, were transmitted orally for centuries before they were written down. After sites of major biblical significance, the lost ancient city of Troy was perhaps the most vigorously pursued archaeological target.

The search for Troy and the eventual excavation of the site in Hisarlik was, at its root, founded on a literal belief in the *Iliad*. The legendary blind poet Homer is credited with composing this work of epic poetry some four hundred years after the events it describes. According to the story, after the Trojan prince Paris ran away with Helen, the wife of Menelaus, the king of Sparta, Menelaus convinced his brother Agamemnon, king in Mycenae, to lead an army of Greeks against Troy in retribution. The epic is full of wild tales of incredible heroism, intervention by various gods, goddesses, and fantastic creatures, and descriptions of a Troy that, as excavation soon revealed, was improbably large and sophisticated. In archaeology's infancy, debate raged over whether Homer's *Iliad* was entirely fictional or founded at some level in historical fact. That debate continues among some scholars today.

The tycoon-turned-amateur-archaeologist Heinrich Schliemann, however, firmly believed in the historical accuracy of Homer's description of the Trojan War. In 1868, Schliemann was excavating in Bunarbashi, Turkey, in a fruitless search for the ancient city of Troy, when he heard of an excavation at nearby Hisarlik. In his journal entry of August 14, 1868, Schliemann noted that he had met "the famous Archaeologist Frank Calvert, who thinks, as I do, that the Homeric Troad was nowhere else but at Hessarlik [*sic*]." Fifteen years before this meeting with Schliemann, Calvert had proposed that Hisarlik was a non-natural mound, with "the ruins and debris of temples and palaces which succeeded each other over long centuries"—in other words, a tell. Calvert managed to acquire ownership of half the hill, and he spent years working to excavate it. He eventually exhausted his resources and had to collaborate with the wealthy Schliemann to continue the exploration (Allen 1999).

Though Schliemann originally noted Calvert's role, he later claimed that Calvert had nothing to do with his inspiration. Eventually, because of Schliemann's greater wealth and notoriety, Calvert's importance in the early archaeology of Troy was completely overshadowed. Although, in his own words, Calvert "had never given much thought to the consequences of the idea," modern scholars increasingly acknowledge that Schliemann owes credit to Calvert for the discovery of Troy (S. H. Allen 1996, 1999).

Schliemann so firmly believed in Homer's account that he may have repeatedly altered the archaeological evidence to support the events in the epic. Nor did he limit his investigations to Troy. Ultimately unsatisfied with the answers yielded there, Schliemann turned to Mycenae, where he excavated what he claimed to be the grave of Agamemnon himself.

Unfortunately, the Schliemann excavations may have forever obscured many of the answers he sought to uncover. Working as he did in the infancy of archaeology, his techniques were crude in the extreme. He identified one layer, known as Troy II, which he thought represented Homer's Troy. He hired unskilled laborers to help him dig down to that layer, discarding the layers of evidence above it. Confused by the complexity of the site and crippled by his own preconceptions (and his habit of playing fast and loose with details), he is widely thought to have fabricated evidence and falsely reported his results. Thus, even in instances where he carefully reported relative positions for his finds, his records are suspect.

Schliemann and his more careful colleague, Wilhelm Dörpfeld, excavated much of the mound where Troy had once stood, reconstructing large swaths of Troy II and dumping the overlying debris off the edges of the mound. After Schliemann's death, Dörpfeld continued excavations and eventually managed to make some sense of what remained of the nine basic habitation layers at Troy. He proposed that Troy VI, not the older and lower Troy II, was Homer's Troy. He went on to excavate some sections of that younger city's wall, which Schliemann had dismissed as Roman work and had partially demolished. However, despite his more systematic approach to excavation, Dörpfeld was still working largely in the dark, without the advantage of the carefully catalogued pottery styles that archaeologists use today to assign dates to habitation layers.

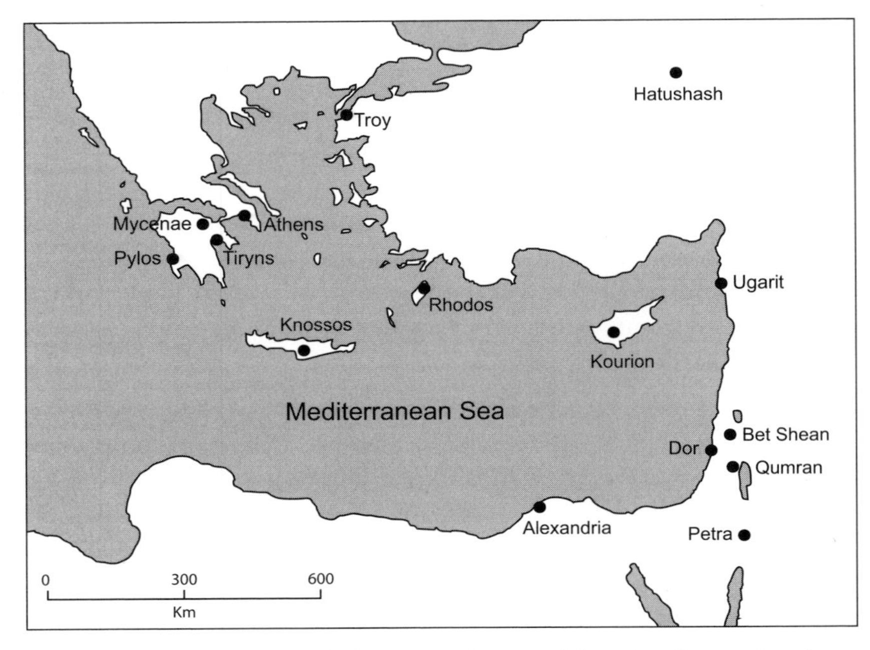

Figure 3.1 A map showing the locations of many of the ruins discussed in the following chapters.

Together, Schliemann and Dörpfeld left little undisturbed at Troy for later archaeologists to examine with more modern methods. Where modern excavation has been possible, there is evidence that earthquakes may have caused some of the destruction at Troy. This would not be surprising; as I discuss in chapter 8, the site identified with Troy is in an area beset by earthquakes today, with earthquakes of magnitude 7 or greater expected every three hundred years or so (Galanopoulos, 1968). In fact, a magnitude 7.4 quake shook the region in 1912, ironically destroying the house of the original excavator, Frank Calvert, four years after his death (Allen 1999, 242). Clearly, however, a great deal of the evidence of earthquake destruction—evidence that would be useful to both earth scientists and archaeologists today—Schliemann dumped off the edge of the mound.

Thus the obsession of many early archaeologists with quasi-historical accounts, such as those found in the *Iliad* and parts of the Bible, has obliterated centuries of unrecoverable knowledge. Even ostensibly historical accounts of ancient earthquakes, written by contemporaries of the events, can bear the taint of cultural, political, or religious bias. This should not surprise us; earthquakes, after all, are emotionally charged events even today, when we understand what causes them and where they are likely to strike. In the past, when people knew of no natural reason for the earth to shake, how could they avoid assigning great significance to this seeming reversal of the natural order, this shifting of terra firma? Although there are tantalizing exceptions, most ancient humans saw the hand of God in any movements of the earth's foundations.

THE WRATH OF GOD

The belief that one or more supernatural beings were responsible for natural cataclysms such as earthquakes, floods, volcanic eruptions, and hurricanes was intrinsic to every known society. Today rationalists often scoff at the notion, as does David Webster (2002), an anthropologist at Penn State University: "One can imagine that unpredictable catastrophic events like these caused a collective psychological panic of 'the gods are angry' kind, even in the absence of serious damage. To me such suppositions are fanciful to the point of absurdity." Nevertheless, the recurring response of societies to large historical earthquakes has been a search for their meaning and an attempt to understand natural disasters in general. Even today, when the natural causes of earthquakes are well known, many people cannot resist the urge to find greater meaning in natural calamity:

> Such occurrences are, on the one hand, natural phenomena like hurricanes, flood, and tornadoes. On the other hand, however, such explanations, while rational, fail to provide us with an answer as to the ultimate cause. Why should the earth's crust not remain completely static and stable? Why do some earthquakes, such as those mentioned in the Bible, occur with such pinpoint accuracy and timing so as to defy human explanation? Why would a loving God allow such disasters to happen? (Fast 1997)

The author of that passage goes on to propose that earthquakes are the result of man's general rebellion against God.

Most biblical references to earthquakes suggest that the calamities are more directly the result of God's pleasure or displeasure with man. Whether the hand of God was sending help or dealing retribution depended on the results of the quake, that is, who was hurt and who benefited. The following account of God aiding the people of King Saul in their battle against the Philistines at Michmash, ca. 1020 BC, appears in 1 Samuel 14:15–23:

> And there was trembling in the host, in the field, and among all the people: the garrison, and the spoilers, they also trembled, and the earth quaked: so it was a very great trembling . . . and, behold, the multitude melted away, and they went on beating down one another. And there was a very great discomfiture. . . . So the LORD saved Israel that day.

God could also send earthquakes as punishment for some shortcoming; even the famous story of the destruction of Sodom and Gomorrah may describe an earthquake, where the Lord "overthrew those cities, and all the plain, and all the inhabitants of the cities, and that which grew upon the ground" (Genesis 20:24–25). So buried in antiquity is this story that no record exists of where these cities were located, although one possible site is the plain north of the Dead Sea, east of Jericho. Accompanying the description is an account of fire and brimstone raining from heaven, and smoke rising from the whole countryside; this could be simple poetic elaboration, but it could describe fires that started during the earthquake or the rising clouds of dust that always accompany earthquakes in such dry areas.

The religious language and ambiguity in nearly every biblical earthquake account may make us uncomfortable, when what we desire is a dispassionate, scientific account. However, our discomfort should not make us disregard these references as completely unreliable. We should not be surprised that the people affected by an earthquake, just like people today, try to find a way to give it some higher meaning in their lives.

EARTHQUAKE MYTHS FROM OTHER TRADITIONS

Other vivid earthquake legends come from outside the Judeo-Christian tradition. The folklore of earthquake-prone Japan blamed earthquakes on the legendary creature Namazu, a great catfish that lived beneath the earth and supported it (Figure 3.2). According to the legend, the god Kashima kept the monstrous catfish subdued by holding a great stone upon its head. Whenever Kashima was not attentive to his task, Namazu flipped its feelers or tail, causing, respectively, small or large earthquakes.

The ancient Greeks blamed the sea deity Poseidon for causing earthquakes. Referred to in the literary works of Homer and Hesiod as *enosichthon,* meaning "shaker of the earth," Poseidon was credited with a host of disturbing actions, from earthquakes to violent storms at sea (Figure 3.3). It seems particularly appropriate that the Greeks would attribute earthquakes to the sea god, for tsunamis—giant, destructive ocean waves triggered by sudden motion of the sea floor—were one of the principal earthquake hazards in the island cultures of Greece.

It is interesting to note that Poseidon is also commonly linked to horses in Greek iconology. This association has led historians to speculate that the well-known story of the Trojan horse could be a heavily veiled reference to an earthquake in Troy. Michael Wood (1985) mentions this possibility when discussing evidence of Trojan earthquakes, but he dismisses it as intriguing but far-fetched.

Ancient Greek texts do more explicitly link Poseidon to some specific quakes. The historic Sparta earthquake, in 469–464 BC, for example, was interpreted as a punishment for breaching certain rules of the god's sanctuary. Thucydides (1910, 60.128) explains: "The Lacedaemonians [ancient Spartans] had once raised up some Helot suppliants from the temple of Poseidon at Taenarus, led them away and slain them; for which they believe the great earthquake at Sparta to have been a retribution."

Archaeological evidence in Linear B tablets unearthed by Carl Blegen (Blegen and Rawson 1966) at Pylos shows that worship

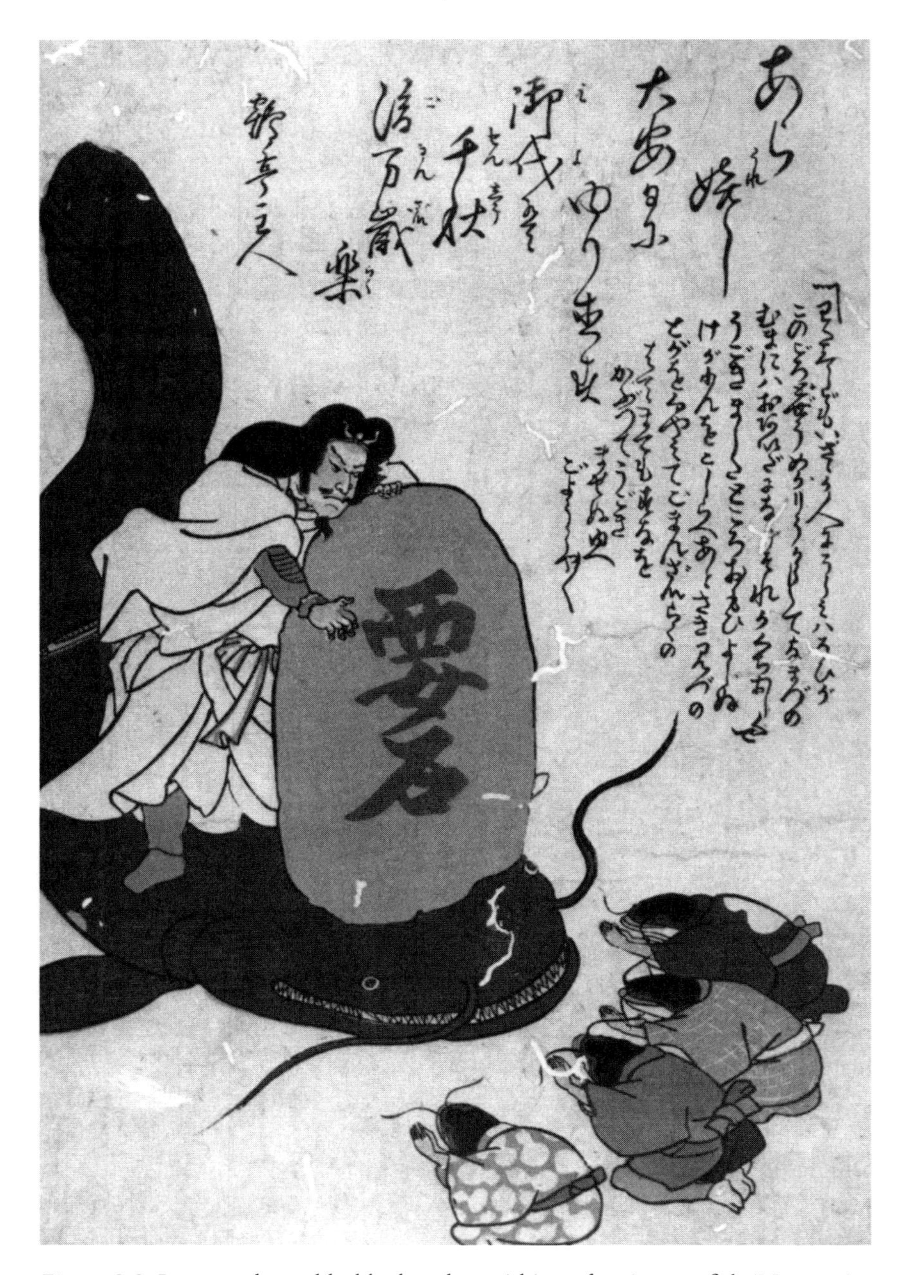

Figure 3.2 Japanese legend holds that the writhing of a giant catfish (*Namazu*) is responsible for earthquakes. This painting shows the god Kashima pinning the catfish beneath a huge stone to keep it still (from Photo Archive Library Earthquake Research Institute/University of Tokyo).

Figure 3.3 The ancient Greeks believed that Poseidon, the God of the Sea, also controlled earthquakes. This engraving by Matthys Pool appeared as the frontispiece of Count Luigi Ferdinando Marsili's *Histoire Physique de la Mer* (*Physical History of the Sea*), published in 1725 (Science Museum/Science & Society Picture Library).

of Poseidon dates back at least to the Mycenaean era. Blood offerings to the god were frequent; Homer's *Odyssey* (1996, 3:5–8) describes a scene of extravagant animal sacrifice by King Nestor and his people: ". . . sacrificing sleek black bulls to Poseidon, god of the sea-blue mane who shakes the earth. They sat in nine divisions, each five hundred strong, each division offering up nine bulls, and while the people tasted the innards, burned the thighbones for the god."

The ancient Romans adopted the Greek view of earthquakes along with the rest of the Greek pantheon. The Roman town of Pompeii, which is most famous for another natural disaster—the eruption of Vesuvius in AD 79—was also prone to severe earthquakes. In fact, in AD 62 or 63—scholars disagree (see Guidoboni et al. 1994)—Pompeii was struck by an earthquake so devastating that the emperor Nero himself came to assess the damage. The emperor recommended that all the residents abandon the city, but not all of them heeded his advice. Those who remained attempted major restorations, without the financial backing of Rome.

The recovery failed. Pompeians struggled to rebuild for seventeen years, until, on August 24, AD 79, the eruption of Mt. Vesuvius buried many of them, along with their handiwork. Thus, almost twenty centuries later, we have unexpected evidence not only of the eruption but also of the city's response to the earthquake seventeen years earlier.

One resident, a banker named Lucius Caecilius Jucundus, memorialized the earthquake with two marble reliefs flanking his household shrine. The artwork, though relatively simple, shows the effects of ground motion during an earthquake. One side shows the dramatic tilting and falling of the temple of Jupiter and other monuments; on the other side, two yoked donkeys are galloping for safety, as the Vesuvian Gate near the aqueduct collapses toward them. Also depicted is a scene of animal sacrifice, reproduced in Figure 3.4. According to Guidoboni et al. (1994, 200), "this represents an offering made to the gods by those people of Pompeii who survived the earthquake . . . an expiatory sacrifice being carried out after the disaster."

Figure 3.4 This relief is from the house of Lucius Caecilius Jucundus, one of the few buildings in Pompeii that were restored after the earthquake of AD 62. On the right is an animal sacrifice, presumably made to appease the gods after the earthquake; on the left, the temples and buildings are shown tilted at alarming angles (Guidoboni et al. 1994, *Catalogue of Ancient Earthquakes in the Mediterranean Area up to the 10th Century*).

Animal sacrifice was common in Greco-Roman culture, but, in unusually terrible catastrophes, there was sometimes human sacrifice as well. Many societies, ancient and not so ancient, have participated in this practice during periods of high seismic activity, further evidence of their conviction that earthquakes came from the gods, perhaps as punishment for some human offense.

One of the most ancient records of ritual human sacrifice that seems to be earthquake-related dates back to 1700 BC, in the Minoan temple of Anemospilia near Knossos, in Crete. We know of this sacrifice because the victim on the altar and three temple functionaries were buried when their temple collapsed on them, apparently shortly after the victim died. This human sacrifice seems to have been an unusual event, since Minoan art shows the usual victims of ritual slayings to be bulls, not men. The forensic evidence from this archaeological find is fascinating, and I describe it in more detail in chapter 5.

THE NEW WORLD

Sacrifices to the earthquake gods were not limited to the Old World, nor were they restricted to the distant past, but the lack of a written record in the Americas makes it difficult to sort out the history of American reactions to earthquakes. We do know that the entire western coastlines of North and South America are extremely seismically active and have been for millions of years, so the ancient people living there must have been quite familiar with earthquakes.

The nomadic people who dominated most of North America in the distant past were probably affected only slightly by any but the greatest earthquakes. Their lightweight construction materials made their homes neither particularly dangerous in earthquakes nor particularly difficult to rebuild. In contrast, the early urban cultures of the Southwestern United States, Mexico, and Central and South America were much more likely to suffer devastation when earthquakes struck.

One possible example is the abrupt collapse of the ancient metropolis of Teotihuacán around 600 AD. Archaeological evidence indicates that Teotihuacán was the largest city in the preindustrial phase of the Americas and the sixth largest in the world at the time, with approximately 120,000 to 200,000 inhabitants. There is consensus that sometime between 700 and 750 AD the city suddenly collapsed, but scholars disagree on the cause (Coe 1962, 105–106; Angulo 1996, 14; and Webster 2002; for additional references, see Tainter 1988; and Manzanilla 2003). There is evidence of looting and burning at Teotihuacán, but as Andrew Coe (1998, 36) wrote:

> They [archaeologists] do not know if this was caused by an internal rebellion or an external attack, but the level of destruction was remarkable. Thousands were killed; their skeletons were trapped beneath the rubble. Those who survived this holocaust gradually drifted away from the city, and a new, less culturally advanced people entered the valley and began to camp out in the ruins. The greatness of Teotihuacán would now only live in memory.

This pattern of squatters amid the ruins of grandeur is quite common following earthquakes. Linda Manzanilla (2003) described widespread "dismantling" of structures and examples of "killed" pottery, figurines, and other artifacts found strewn on the floor, suggesting this was the result of an internal "ritual destruction" by the subjects against the ruling caste of the city. This area, however, is very active seismically. An earthquake striking this site does not rule out a concurrent internal or external rebellion or uprising of enslaved populations at Teotihuacán. However, an earthquake might very well have been a spark igniting the tension inherent in the caste system. Teotihuacán was ruled by a priestly caste, and the people probably expected their gods to protect them. Failure to control an earthquake would have undermined the priests' power and may have stirred up a rebellion.

Unfortunately, there are no earthquake accounts to help us outline past seismic activity here. However, several suggestive features have been observed. The offset of the staircase in the Central Patio of the East Plaza complex at Teotihuacán shows that the structure was rebuilt in a slightly different location than the one below it, which had been destroyed, perhaps by an earthquake (Figure 3.5); closer investigation reveals that the staircase was twice rebuilt over earlier staircases. The shape of the main avenue's massive Pyramid of the Moon and Sun would have been resistant to earthquakes, and so the main bulk of these structures remain intact. The rubble atop the Pyramid of the Moon implies, however, that the structure was damaged at some time. We may never know whether it was "dismantled" intentionally or simply destroyed by the action of time and climate. Destruction by earthquake is certainly the simplest explanation and is consistent with both the pattern of destruction and the area's tectonic setting.

The Aztecs were fully aware of the occurrence of earthquakes and gave them particular significance in their religious life. The complicated Aztec religion held that the universe had been created five times, and that each of the four preceding worlds, known as "suns," had been destroyed by its own particular catastrophe. They believed that the first four suns were destroyed, in turn, by

Figure 3.5 The steps of the Central Patio of the East Plaza complex at Teotihuacán, which was rebuilt at least twice, more or less on top of earlier collapse. The collapse most likely was caused by an earthquake (Photo: Enrique Franco Torrijos).

jaguars, storms, a rain of fire, and a flood. The Aztecs believed that the fifth sun, their own, would be destroyed by an earthquake.

Such a worldview must have contributed cataclysmic overtones to any major seismic event. There is evidence of regular human sacrifice in Aztec culture, sometimes specifically associated with earthquakes. However, we have no written records of sacrifices that predate the arrival of the Spanish, and we can only speculate whether earlier sacrifices had similar causes. Patrick Tierney (1989) provides an interesting overview of the subject in his book, *The Highest Altar: Unveiling the Mystery of Human Sacrifice.*

Tierney points out that the most recent American example of human sacrifice in response to an earthquake (and its accompanying tsunami) was after the largest earthquake ever recorded with modern instruments, the magnitude 9.5 Chilean earthquake of May 22, 1960. After this earthquake, a group of Mapuche Indians in the Pacific coastal town of Lago Budi sacrificed a five-year-old boy by cutting off his limbs and leaving him for the waves in hope of satisfying the malevolent forces that had caused the destruction. The perpetrators of this atrocity, who included the boy's grandfather, were tried in Chilean courts for their actions, but after serving two years in jail awaiting their trial they were released. Most modern Chileans, and many of the Mapuche, were horrified by this event. Still, more than two decades later, Tierny interviewed an eighty-five-year-old Mapuche elder, who bemoaned the passing of human sacrifice. In her opinion, when they had followed their tradition of sacrificing orphans in the past, there were fewer quakes and tsunamis.

Although this reaction to a modern earthquake is, I think, an atavism almost too gruesome to believe, people have been responding to catastrophes in this way for millennia. Gunnar Heinsohn (1998) produced a study of Bronze Age blood sacrifice, and although some of his theories about the Bronze Age are suspect—including his proposal to radically alter the chronology of world history—he appears to be credible in his assessment of why blood sacrifices may have come into existence around the same time people began gathering in cities:

> The community that participates in the great play is able to release in the act of killing the aggression born out of helplessness with regard to the catastrophe. The slaughter of the sacrificial victim is both the conclusion and the cathartic climax of the ceremony. Men liberate themselves in this bloody act from the fury that until then had been turned inward and that had caused them helpless numbness, psychosomatic suffering or aggressiveness which endangered their fellow men.

THE SEEDS OF SEISMOLOGICAL THOUGHT

This common brutal reaction to calamity in antiquity was not universal, however. The view of earthquakes as mystical phenomena existed in parallel with more pragmatic interpretations, even in the distant past. Throughout history, even those who attributed earthquakes to God or gods also struggled to understand the natural forces that could shake the ground with such violence. An early English book on historical earthquakes, for example, includes the following observation:

> An Earthquake is a shaking of the Earth, occasioned by Wind and Exhalations inclosed within the Caves and Bowels of the Earth, which can find no passage, or at least none long enough to discharge themselves, and therefore braking forth with force and violence, it sometimes shaketh the Earth. . . .
>
> But though we have given some account of the Natural Causes of Earthquakes, yet it is very apparent that many have been supernatural, and caused by the immediate hand of God, of which we find several instances in holy Scripture, that we might dread and tremble before the Almighty who needs neither Vapours nor Exhalations to Execute his Vengeance upon incorrigible Offenders. (R. B. 1694)

We cannot know, of course, whether this duality reflects the author's real theories or whether he added the bit about divine causes to avoid censure by the church. However, this mixed viewpoint can be found even in the Bible itself. Not surprisingly, the Bible usually assumes that earthquakes are deliberately sent by God, either as punishment or as corroborating evidence of a larger miraculous event. A more equivocal view, however, is expressed in 1 Kings 19:11–12, where the prophet Elijah, wishing to die after being rebuffed and rejected by the people of Israel, stands atop Mt. Horev:

> And behold, the LORD passed by, and a great and strong wind rent the mountains, and brake in pieces the rocks before the LORD; but the LORD was not in the wind: and after the wind an earthquake; but the LORD was not in the earthquake: And after the earthquake a fire; but the LORD was not in the fire: and after the fire a still small voice.

I interpret this to mean that the writer of 1 Kings believed that wind, fire, and earthquakes, although frightening events, were natural occurrences; God was the whisper in the stillness.

Other writings contemporaneous with parts of the Bible make it clear that a secular view of earthquakes was not unheard of in Jewish society. According to the Jewish historian Flavius Josephus (1991a [ca. AD 75, translated in 1737]), Herod the Great, in 31 BC, made the following speech to reassure his troops after a devastating earthquake:

> Do not disturb yourselves at the quaking of inanimate creatures [nature], nor do not imagine that this earthquake is a sign of another calamity [to come]; for such affections of the elements are according to the course of nature; nor does it import anything further to men, than what mischief it does immediately of itself. Perhaps, there may come some short sign beforehand in the case of pestilence, and famines, and earthquakes; but these calamities themselves have their force limited by themselves.

In reality, these words may not be Herod's at all. Historians in the time of Josephus commonly enlivened their accounts with fictional speeches attributed to key figures. Still, whether these words are indeed Herod's or Josephus wrote them after the fact, they describe this earthquake, which I elaborate on in chapter 6, as not only a natural phenomenon but one that someday may be predicted, perhaps by some sort of precursor. Most important, the attribution has nothing to do with God.

There is an even earlier example from ancient Nineveh in an Assyrian letter written in one of the outlying towns of the empire some time in the eleventh century BC (Ambraseys and Melville 1982; after Thompson 1937):

> On 21 Elul [Syrian calendar; see Grumel 1958, 174] an earthquake took place. All the back part of the town is down; all the wall at the back of the town is preserved [except] 30 1/2 cubits therefrom being strewn and fallen on the near-side of the town. All the temple is down. . . . Let the Chief [architect] come and inspect.

There is no mention or implication here of the wrath of God, only concerns over repairing damage and restoring order in the face of a natural disaster.

Natural causes for earthquakes were also favored by the majority of ancient Greek philosophers, including Thales, Anaximander, Anaximenes, Parmenides, Anaxagoras, Zeno, Democritus, Plato, Aristotle, and Epicurus, though they rarely agreed on the specifics. Thales and Democritus favored underground water as the driving force for earthquakes, Anaximenes preferred the collapse of underground cavities, Anaximander and Aristotle favored underground winds or "exhalations," and, finally, Anaxagoras was quite catholic in his causes, favoring, in turn, collapsing cavities, underground fires, and pockets of the ether trapped underground, struggling to rise.

Following in the footsteps of these Greek sources and others, Roman historian and philosopher Seneca, after the Pompeii earthquake, further explored the causes of earthquakes. He mostly favored air as the catalyst, agreeing with Anaximander and Aristotle, but differing with them regarding the mechanism. He was, however, just as adamant that earthquakes were not the result of the supernatural energy of the gods: "It will help also to keep in mind that gods cause none of these things and that neither heaven nor earth is overturned by the wrath of divinities. These phenomena have causes of their own; they do not rage on command but are disturbed by certain defects, just as our bodies are" (Seneca AD 65).

These erratic early attempts at earthquake theory reflect the confusion born of looking at a complex problem with inadequate data. A full understanding of earthquakes requires knowledge of plate tectonics, an idea, of course, that would never have occurred to the Greeks. For they had no way to pinpoint the precise locations of earthquake foci, nor did they have anything like the modern, worldwide observation system necessary for measuring the properties of the earth's interior and tracking the motion of points on the earth's surface. Today scientists have a tremendous advantage: they know that earthquakes happen on faults, that they release stress below the earth's surface, and that there are patterns to their recurrence and strength. Long-term tracking using global positioning system (GPS) satellites has even allowed us to measure the slow movement of the tectonic plates. In the discussions of archaeological sites in the chapters that follow, I explain what we know about earthquakes, how they affect structures, and how that knowledge should affect archaeological approaches in earthquake-prone regions.

The archaeology of sites such as Jericho and Troy, with their layers of destruction and histories of early, crude excavation, are too complex and muddled to serve as starting exhibits of earthquake damage. Instead, I begin with other, less ambiguous sites that illustrate the types of clues earthquakes leave in the archaeological record. Later, I return to discuss Jericho in more detail, along with other sites, including Troy and Mycenae. I also examine how earthquakes have affected societies, counteracting or intensifying social and political problems, and influencing a wide range of societal changes, from the end of the Bronze Age to the beginning of the Age of Reason. Earthquakes have even helped to determine what we can uncover from the past, from the earliest Neanderthal societies to the discovery of the Dead Sea Scrolls.

CHAPTER 4

Clues to Earthquakes in the Archaeological Record

Everything should be as simple as possible, but not simpler.

—attributed to Albert Einstein, source unknown (see Calaprice 2000)

The primary challenge of earthquake archaeology is to determine whether the destruction of a site was caused by an earthquake, by human attack, or by some non-seismic natural cause. In any excavation in a known seismic hazard area, excavators should always be alert to evidence of earthquake damage, even when they know or suspect that human action was responsible for the destruction of the site. Layers above or below the horizon of interest may have been damaged by earthquakes, or the entire site may have been damaged after the fact, scrambling the archaeological record in sometimes subtle ways. Knowing the kinds of damage that occur in earthquakes can help archaeologists make sense of the sometimes confusing evidence that major quakes leave behind.

GROUND DISTURBANCE

One of the most definitive signs of earthquake damage is displacement of the ground surface. This effect cannot be mimicked by the actions of invading armies, and it is readily preserved in the archaeological record. Whereas geologists know of several types of seismic ground disturbance, only fault motion is familiar to most archaeologists.

Fault motion, as described in chapter 2, can result in vertical or horizontal displacement of the ground surface on one side of the fault relative to the surface on the other side. Sometimes the motion causes tilting of recognizable structures that should be horizontal or vertical, such as walls, or, say, the pebble floor at Ubediyeh mentioned in chapter 2. Vertical displacement that creates a step-like discontinuity can be readily apparent when an affected site exhibits easily distinguished habitation levels. This sort of evidence is what Robert Drews was looking for when he wrote the words quoted in chapter 2. However, finding a layer that is offset by vertical displacement on a fault does not necessarily imply that an earthquake occurred while that layer was inhabited. Each layer accumulates on older layers beneath it—a principle known in geology as *superposition*—and an earthquake that breaks the ground surface also breaks all the layers beneath that were deposited (and abandoned) before the earthquake. Only by dating a layer that succeeds the earthquake—one that overlies the fault and is not broken by it—can scientists determine a date *before which* the earthquake must have occurred.

The mainly horizontal motion that occurs on transform faults leaves less obvious markers in the archaeological record, except where it cuts through linear surface features such as walls, streets, or aqueducts. Walls bisected by active faults are displaced across the fault, and this displacement can be used to draw important conclusions about cumulative motion on the fault, especially when the construction dates are known.

Metzad Ateret

When we are fortunate enough to find evidence of faulting in an archaeological site, it can be remarkably informative both to archaeologists and to seismologists seeking to improve our knowledge of

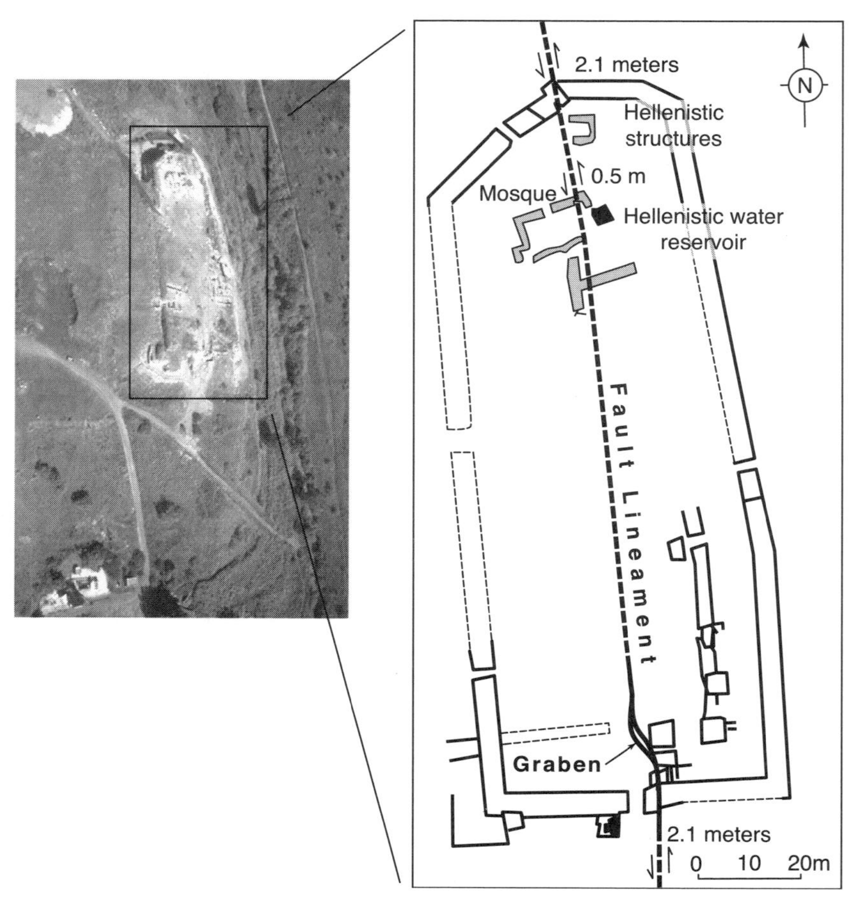

Figure 4.1 The Crusader fortress of Metzad Ateret in northern Israel (Courtesy Ateret Research Group).

seismic risks in the surrounding region. An excellent example of such a lucky synergy is the excavation and geologic trenching done in northern Israel (Ellenblum et al. 1998), at the crusader fortress known variously as Metzad Ateret, Vadum Iacob, and Le Chastellet (Figure 4.1).

The Dead Sea Transform fault bisects the walls of this fortress, offsetting the oldest walls by more than 2 meters across the fault (Figure 4.2). However, since earthquakes occur repeatedly

Figure 4.2 A photo of the displaced outer wall of Metzad Ateret, where the line of the wall has been offset by 2.1 m in several earthquakes. The dotted line follows a single course of bricks, and the solid lines illustrate the offset between the wall on one side of the fault zone relative to the wall on the other side.

on the same faults, the offset itself is not enough to tell what happened here.

What seismologists want to know about faults is twofold: How large are the earthquakes on a given fault, and how often do they recur? Knowing these parameters, seismologists can draw conclusions about the likelihood of large earthquakes in the near future on the same faults. What we really want to know is where and when earthquakes will strike, but because there has been little progress toward actual earthquake forecasting, we have to content ourselves with making our statistical risk analyses as accurate as we can.

Geologists and archaeologists at Metzad Ateret have carried out collaborative trenching excavations, where a trench is dug across the active fault. By cutting through the layers that were broken in the earthquake or earthquakes and exposing them in the walls of a trench, geologists can tell how many events are represented in the layers. This is because, even though the same fault slips repeatedly, the cracks that reach the surface tend to follow slightly different paths each time. We may see, for example, that a lower layer was broken twice, but only one of the cracks also extends into the layer above and the other terminates abruptly at the newer layer. This tells us that two earthquakes affected the lower layer, one before the upper layer was deposited and one afterward.

This is exactly what the researchers at Metzad Ateret discovered. Researchers determined from archaeological data that one major earthquake, with around 1.5 meters of offset, occurred shortly after the conquest of the crusader fortress; the seismic action occurred after the stones from the wall had been thrown down but before the mud-brick fill had eroded from above and buried the invasion layer. Historical documents date the invasion to AD 1179. At least one other earthquake occurred later, displacing by more than half a meter the walls of a mosque built atop the ruins of the fortress. Pottery evidence dates the mosque damage to somewhere between 1517 and 1917, during the Ottoman period. So, although the total movement on the fault since the crusader fortress was built is 2.1 meters, we know that this displacement was divided between at least two different earthquakes.

Qumran

Another case of ground displacement in archaeological ruins can be seen at the archaeological excavation of Khirbet Qumran, an enigmatic site near the Dead Sea that became famous when it was linked to the finding of the Dead Sea Scrolls in nearby caves. Figure 4.3 shows the steps leading down into a cistern at the Qumran site. The crack that offsets the eastern side of the steps from the western side (by around 50 cm) was caused by an earthquake. The fault line runs through this cistern and an adjacent one, and continues through the site, damaging several of the buildings there. This damage was most likely caused by the Judean earthquake of 31 BC. The precise date of this earthquake, as well as almost

Figure 4.3 The cistern at Qumran near the Dead Sea. The steps leading down into the cistern have been broken by a fault, possibly by an earthquake in 31 BC. Note how each step was displaced across the fault.

everything else about Qumran, is disputed, but that it occurred sometime near 31 BC is well documented. I am convinced that the 31 BC earthquake did in fact devastate Qumran and the nearby caves, and I will return to describe the supporting evidence and discuss the implications of this in chapter 6.

Unfortunately, many of the clues that would allow us to narrow the timing of this earthquake damage were discarded with the material that was removed in the excavation. Most of the site was excavated by Roland de Vaux, who died without publishing a final excavation report. Therefore, we do not know how carefully he examined the traces of the fault in the debris layers he removed. He interpreted the fault as having broken during the 31 BC earthquake, but modern scholars have proposed alternate explanations, including a later earthquake after the site was abandoned. The date could have been more accurately assessed had any portion of the sediments above the broken steps been left in place.

Of course, earthquakes are not the only natural phenomena that cause cracks to appear in the ground or in structures. Landslides or subsidence of fill beneath a building can cause walls to crack or fall. This is a complicated issue, because both subsidence and landslides are frequent occurrences during earthquakes, but they can also be caused by periods of heavy rain, flooding, removal of vegetative cover on a slope, or simply bad building practices. To distinguish between ground motion from earthquakes and motion caused by these other factors, we must carefully examine the topography of the affected region as well as the extent of the damage. For example, in a landslide, we would expect only steep slopes to be affected, and only in isolated instances. Landslides throughout a site would indicate a possible earthquake, where shaking weakened a broad area of sediments perhaps already saturated with water. Walls on flat ground would not be affected. Likewise, if ground motion were caused by settling of fill due to bad building practices, we would expect the damage to be extremely localized to the spot or spots that subsided.

The final type of ground disturbance from earthquakes, to my knowledge, has not yet been recognized in any archaeological excavation, but it has been uncovered by geologists in trenches dug across faults. It occurs when wet sand or clay beneath the surface is violently shaken and liquefies, erupting to the surface in features called "sand boils" or "sand volcanoes." In excavations, the path that the liquefied sand took to the surface is evident as a "liquefaction column," or a column-shaped tube of sand that cuts through sediments between the deep source of the sand and the surface. If this feature were found in an archaeological excavation, it would be unequivocal evidence of an earthquake, and the age of the layer where it reached the surface would provide a means of fixing the earliest possible date for the event.

FALLEN COLUMNS

Permanent ground distortion, although compelling when we discover it in excavations, is not the cause of most earthquake damage.

As discussed in chapter 2, most of the destruction from earthquakes comes from transient shaking away from the culprit fault during the event. When the earthquake is over, the only evidence of the shaking (in the absence of modern seismograph records) is the collapse it leaves behind. Distinguishing earthquake collapse from other types, however, can be exceedingly difficult a thousand years or more after the fact.

Fallen columns offer some of the best and simplest evidence for past earthquakes, and Roman columns are especially instructive. Roman roofs, even those of imposing stone buildings, were typically made of wooden beams and boards covered with terra cotta tiles. This design offers little resistance to the horizontal ground motion that causes most damage from earthquakes. As a result, the columns in many ancient Roman structures were essentially freestanding, with nothing to hamper their falling in an earthquake.

An attacking army, by applying a great deal of manpower or ingenious leverage, was capable of toppling the columns of a building as part of laying waste to a conquered city. One proposed mechanism for ancient destruction of colonnades involves wedging great timbers between columns and then drenching them until they swell, thereby forcing the columns apart. However, as with other methods using human force, this one produces a chaotic collapse very different from what we see in a major earthquake, and, for all human techniques, the larger the columns, the more difficult they are to shift.

In earthquakes, on the other hand, the mass of enormous columns increases the tendency for them to shift and fall. When we search for clues to earthquake damage in fallen colonnades, we look for a common direction of fall, either in all the columns in a single building or in all the buildings in a single site. The orderly alignment of fallen columns, where an entire row of heavy columns is found toppled in the same direction, is unlikely if the buildings simply crumbled with time or foes forced them apart.

The sudden ground motion of an earthquake, however, produces just such an orderly pattern, one that we often find in archaeological excavations. At the start of an earthquake, the ground, along with the column bases resting on it, shifts suddenly in one direction.

Because of inertia, the tops of the columns tend to be left behind. If the motion is severe enough, the columns may topple entirely, all in the direction opposite to the initial strong ground motion.

A major earthquake that occurred in the Holy Land around AD 749 (Tsafrir and Foerster 1992), causing destruction to such places as Capernaum, Susita, Jerash, Tiberias, Gadara, Pella, Scythopolis, Jerusalem, and Philadelphia, left behind several particularly good examples of collapsed columns. The excavated ruins of two towns—Susita on the Arabian Plate and Bet Shean on the Mediterranean Plate (Figure 4.4)—tell an interesting story.

Susita

Susita is a ruined city on a steep, isolated hill, overlooking the Sea of Galilee from the East. It was a thriving community during Roman and Byzantine times but began to decline in importance when the Arabs gained control of the region. The final blow, however, was struck not by the Arabs but by the earthquake of AD 749, which destroyed the city's large temple, and much else besides. Among the massive piles of rubble scattered about the hilltop are nine stone columns, each weighing about 15 tons, which all fell in the direction opposite to the ground motion caused by the quake and now lie parallel to one another in the dust (Figure 4.5). This is what we would expect from a strong earthquake. The marble and granite columns were similar in weight, height, and construction, and since they were located close together, they experienced essentially the same ground motion during the earthquake. The effect would have been much like pulling a tablecloth from under a row of tall champagne flutes; as the ground accelerated, the inertia of the tall, massive columns resisted the motion. With their bases pulled from beneath them, the columns were overbalanced and toppled.

Looking more closely at the columns in Figure 4.5, we see that the columns all lie with their axes pointing southwest. However, if we draw lines that connect the center of each stone base with the center of the column that once sat upon it, most of those lines point

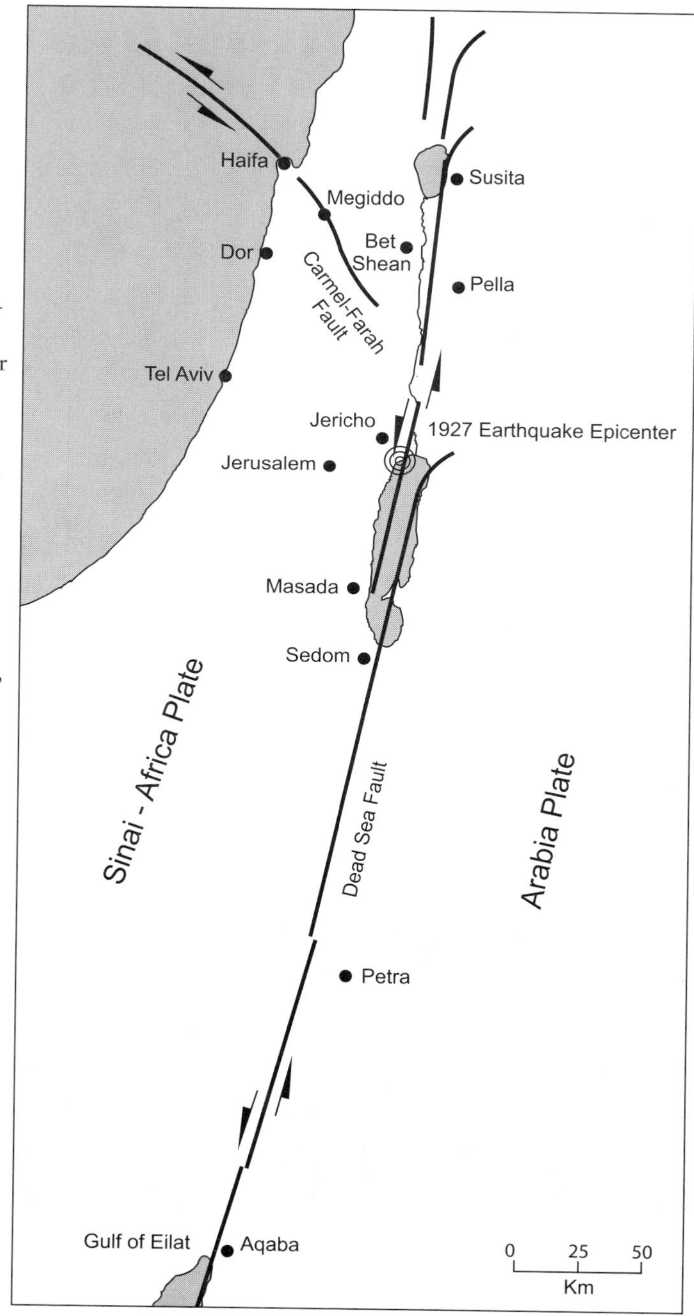

Figure 4.4 A location map of the Dead Sea Transform region, showing the locations of major faults and the epicenter of the 1927 Jericho earthquake. This event was the largest so far recorded in the region with modern seismographs, and is discussed in more detail in chapter 7.

Figure 4.5 Fallen columns in the Byzantine city of Susita on the Sea of Galilee, which was destroyed in an earthquake in AD 749. Note how the columns are parallel to one another.

south*east*. This is because, in an earthquake, the ground moves back and forth and continues to move for some time. The columns may have initially been bucked off their bases to the southeast and begun to tip. The continued ground motion shifted the direction of fall, with the final line of the columns pointing southwest. Alternatively, they may first have fallen to the southwest and then all rolled with the continued ground motion; this scenario, however, is less likely, since, on the irregular ground surface, which would have been littered with earthquake debris, each column probably would have rolled a different distance in a different direction, resulting in a more chaotic arrangement of the columns. The southward component of motion, at least, is expected; the long-term motion of the Arabian Plate is to the north, and we expect the initial northward jump of the plate—the jerked tablecloth—to result in the columns toppling to the south.

Bet Shean

Some 45 kilometers south of Susita lie the ruins of ancient Bet Shean, a city mentioned several times in the Bible, as well as in ancient Egyptian sources dating back to Pharaohs Thutmose III (fifteenth century BC) and Ramses III (twelfth century BC). Like Susita, Bet Shean was destroyed by the AD 749 earthquake, but, as the former regional capital, Bet Shean was the more imposing of the two cities. The two intersecting main streets of the city were lined with stone columns, some as tall as fifty feet—columns that now lie fallen in their ranks, still parallel to one another (Figure 4.6).

Excavators were initially attracted to a particular area in this site by what appeared to be the top of a column protruding a foot or so above the surface of the ground. This column "top" turned out to be the tallest remnant of an entire city in earthquake ruins. Excavations of the Byzantine part of the city during the late 1980s and early 1990s uncovered a few primitive stone huts, built among the collapsed buildings. These huts, possibly cobbled together by earthquake survivors for emergency shelter, are the only evidence

Figure 4.6 The ruined columns of Bet Shean, toppled by the AD 749 earthquake.

of habitation after the quake. Otherwise, the city was apparently abandoned.

Most of the columns in Bet Shean fell in nearly the opposite direction from those in Susita. This is to be expected: Bet Shean is not on the Arabian plate, which moves north, but on the Mediterranean plate, which moves south. Therefore, if these columns fell during the first strong ground motion of the earthquake, they should have fallen in the opposite direction from the ones in Susita.

During the excavation, however, some columns in Bet Shean were found in orientations inconsistent with the general pattern. In fact, the foundations of several of the Byzantine colonnades were made of columns lying on their sides, a rather unusual construction technique. Stranger still, these foundation columns were made of foreign granite, not the locally quarried limestone of the colonnades they supported (Figure 4.7). There must have been an earlier destruction here, one complete enough that none of the older construction style remained standing to fall with the Byzantine layer in AD 749.

Figure 4.7 Granite columns at the base of the Byzantine wall in Bet Shean. The granite columns are from a previous building of the Roman occupation, and have been reused in the later construction.

We now know, based on written evidence, that this city was, in fact, destroyed in an earlier earthquake, in AD 363, when Bet Shean was under Roman control. We know that granite columns were imported by the Romans for many of their main cities, rafted down the Nile from Upper Egypt, then brought by ship to Caesarea, and finally hauled overland. Apparently, these columns, now fallen and broken, were reused when Bet Shean was rebuilt following the 363 earthquake. Archaeological evidence reveals that the Byzantine builders used many of the Roman stones and columns as foundations and general building material for their reconstruction of Bet Shean. The granite columns, though perhaps not suited to the style of the new design, nonetheless found good use as part of the foundations for the Byzantine main street colonnade.

A coherent pattern of fallen columns not only provides some of the most direct evidence for earthquakes but can also allow us to say something about the direction of the first ground motion in the earthquake. In this region, where we know the motion expected on the Dead Sea Transform, the motions indicated by these two sets of columns are not surprising. They provide hints, however, that the earthquake probably did occur on the Dead Sea Transform and not on one of the many smaller faults that crisscross the area, having many orientations and fault mechanisms—strike-slip, normal, and reverse.

Modern seismologists use first motions recorded by seismographs located in all directions from the earthquake's epicenter as part of a tool kit to determine the kind of fault mechanism responsible for the shaking. Similarly, by analyzing the direction of column collapse in ancient earthquakes, it is theoretically possible to discover something about ancient earthquake mechanisms. The column collapse patterns in Bet Shean and Susita in the AD 749 earthquake are a crude example of this, showing only the approximate direction of first ground motion, and only at those two sites. With more sites, we could theoretically say much more.

Unfortunately, this theory is of limited value in practice. The first motion must be large enough, in the first place, to knock down the columns, and this occurs only in large earthquakes. In smaller earthquakes, the first motion might only start the columns rocking, and subsequent shaking would wobble them around until they fell in a more chaotic way. Second, the widespread use of free-standing columns in this area lasted only approximately one thousand years, after which wooden roofs were reinforced with heavy stone headers connecting the tops of the columns. Such bracing often complicates the pattern of collapse.

In both cases, collapse itself may still be suggestive, but when the directions of collapse are jumbled, it becomes harder to determine whether earthquakes were even the cause, much less to know what type of earthquake might have been responsible. Furthermore, with post-Byzantine architecture, where columns became tied to non-circular arches and other fortifying structures, buildings became much less vulnerable to moderate and small earthquakes. As a result, many structures from the later periods are still standing, despite having been lightly shaken once or twice. Thus, although the columns of Roman architecture provided an interesting glimpse into ancient earthquake mechanisms, the thousand-year time window they afford, particularly when viewed through the archaeological layers of the intervening millennium, is simply not long enough to give us much insight into the seismic hazards of the area.

Petra

Ancient historians described the city of Petra, in what is now Jordan, as the political and cultural center of the Nabateans. This ancient city sits in the Dead Sea Rift Valley to the east of Wadi Arabah, close to the seismically active Arabian plate boundary. It has a long history of excavations of its more than eight hundred monuments since the Swiss traveler Johann Ludwig Burckhardt stumbled upon it in 1812. Although the city had remained known to local Bedouins through the years, Burckhardt's discovery of it reintroduced it to the wider world.

Today, Petra is best known for its breathtaking tombs and classical facades, carved into the solid rock of its imposing red sandstone cliff faces (Figure 4.8). In its heyday, however, Petra also had numerous freestanding buildings in the open area between the cliff faces, including a paved street lined with columns (the Colonnaded Street) and numerous temples and residential buildings. These freestanding buildings, with columns made of stacked stone disks, were much more vulnerable to earthquake damage than the buildings carved of solid stone, and thus few of them remain standing today.

The massive earthquake of May 19, AD 363—the same one that leveled Roman Bet Shean's imported granite columns—devastated most of Petra as well, along with many other cities in the region. Many of the monuments and buildings in Petra's ruins have been reconstructed since its rediscovery in 1812. So much has been excavated and rebuilt, in fact, that it is often difficult to determine the causes of the original destruction. However, in spots where the ruins are relatively undisturbed, earthquakes have left an unmistakable mark. Though Petra's columns were not monolithic like those in Bet Shean and Susita, they show a remarkably orderly pattern of

Figure 4.8 The famous facade of the Nabatean city of Petra in Jordan, carved out of the sandstone cliffs.

Figure 4.9 Fallen columns in Petra, destroyed in the earthquake of AD 363.

collapse. In many places, columns lie fallen in the streets, each disk leaning against its neighbor, like vertebrae in a still-articulated spine (Figure 4.9).

Petra was hit by several other earthquakes after the 363 event, some causing significant damage. Certain scholars attribute Petra's final destruction to the earthquake of July 9, 551 (Taylor 1993, 17) or to a different one in the eighth century (Browning 1989, 60). However, excavations led by Bernhard Kolb of Basel University in the late 1990s revealed that the damage from the 363 earthquake was extensive, confirming that many buildings were never reoccupied or rebuilt after this earthquake. The remarkable preservation of the collapsed columns certainly indicates that the city never completely recovered after 363.

Chichén Itzá

A final example of fallen columns comes from Mexico, a seismically active place with no written record of pre-colonial earthquakes.

Figure 4.10 The building of the Temple of the Warriors, from the west, with columns along the front and southern sides, located in Chichén Itzá, Mexico. The structure, also known as "Complex of a Thousand Columns," is thought to have been built after AD 849, and the orderly array of fallen columns indicates that it was destroyed by an earthquake some time after that (from Sharer 1994, with permission of Carnegie Institution of Washington: Washington, DC).

Today, the ancient Mayan city of Chichén Itzá is among the most-visited archaeological sites in Mexico. Situated on the Yucatan Peninsula, the site functioned mainly as a religious ceremonial center, with palaces, temples, and pyramids. The site's architecture was probably influenced not by a single culture but by two eras overlapping substantially, but there is little consensus on its dates of occupation and collapse (Sharer 1994).

The key point of interest here is the "Complex of a Thousand Columns," bordering the front and southern side of the monumental Temple of the Warriors. These colonnades are remnants of a huge structure that scholars believe was erected sometime after AD 849. We are fortunate to have a picture of the fallen columns taken before much of the reconstruction began (Figure 4.10). Hundreds of columns were arrayed in an orderly manner, all facing the same direction. These columns were consequently restored to their original bases (Figure 4.11). The photo is intriguing, however, as

Figure 4.11 The Temple of the Warriors at Chichén Itzá, taken from the top of El Castillo, as it is today. It is fortunate that the earlier photograph shown in Figure 4.10 survived, or the information implicit in the pattern of column collapse might have been lost (Photo by Robert Clapp © 1999).

it suggests the possibility, certainly not surprising in light of the seismic hazards in the area, that an earthquake caused the destruction of ancient Chichén Itzá.

WALLS

Like fallen columns, collapsed walls—especially those made of stone or mud brick using ancient techniques—are often clear markers of past earthquakes. A single site in any given earthquake generally has one axis where the shaking is the strongest. Walls oriented perpendicular to this maximum-shaking axis are more likely to collapse in moderate earthquakes, leaving walls that are oriented parallel to the shaking axis intact or less badly damaged. When many similarly oriented walls at a site have fallen in the same direction, particularly when they buried grain, gold, or other valuables in their fall, the action of an army is an unlikely cause. We find such damage in many different sites, including some layers at Troy, Jericho, and Mycenae. The ruins of storerooms at Masada in Israel illustrate this effect better than any other site I know.

Masada

At the end of the Jewish war against the Romans in AD 70, the stronghold of Masada (or Metzada [*Met-zah-DAH*] as it is pronounced in Hebrew), which towers over the Dead Sea in the Judean Desert, became the last free Jewish community in Judea. Stories of its destruction are debated, but the most moving account comes to us from the writings of the Jewish historian Josephus. According to Josephus, around 960 Zealots—men, women, and children—lived on this rock for three years, surrounded by the Roman army. Protected by the natural steep cliffs and the fortifications Herod had added at an earlier time, they managed to fend off the Roman legions for those three years. Intent on victory, the Romans finally erected with enormous effort a 450-foot ramp on the western side of Masada to break into the top of the fortification. A day before the Romans breached the final walls, the defenders committed mass suicide, each man killing his own family before being killed by a friend or killing himself. The conquering Roman army breached the citadel and marched in the next day to find 960 bodies in the stillness of this mountaintop.

Although the Roman siege of AD 70 was responsible for the death of its people, an earthquake, perhaps the one in 363, appears to have caused the final destruction of Masada's remarkable buildings. One feature that made it possible for Masada's people to withstand the Roman siege for as long as they did was their elaborate system of storerooms (Figure 4.12) and huge water cisterns. The storerooms were designed as a series of long, narrow chambers parallel to one another, with walls of unreinforced stone eleven feet high. This type of construction is particularly vulnerable to shaking perpendicular to the long direction of the storerooms, since the long walls have no bracing along most of their length. The shorter walls are braced at every intersection with the long walls, at least until the long walls collapse.

The ruins at Masada show that the storehouse walls apparently collapsed as a single unit, with each wall falling into the storeroom next to it. The rows of stones seen in the photo in Figure 4.13,

Figure 4.12 Aerial view of the Jewish stronghold of Masada, showing the partially restored storerooms and the huge cistern that made it possible for the 960 zealots to withstand the Roman siege for so long.

Figure 4.13 Photo of collapsed storeroom walls in Masada. Note how the rows of masonry are still clearly visible, as if the wall collapsed outward as a unit.

which fortunately was taken before excavation began, are the kind of marker we have come to look for as evidence of an earthquake.

In many places where earthquakes have toppled walls like these, we find a bounty of stored food, a treasure trove of valuables, or even the remains of people trapped beneath the collapsed walls, never to be retrieved after the destruction. In the case of Masada, however, little was found. The inhabitants had been dead possibly for three hundred years, and the invading Roman army had long since removed anything of value.

After the excavation, some of the storerooms were restored, as seen in Figure 4.13, rebuilt by archaeologists. The efforts these archaeological volunteers put into rebuilding this site is impressive, but one must wonder when the next earthquake will simply knock it back down.

Aqaba

Also destroyed by the AD 363 earthquake was the Aqaba Basilica in Jordan, some 200 kilometers south of Masada, where an excavation shed new light on the seismic activity in the area. Field studies he directed starting in 1994 led Thomas Parker of North Carolina State University to conclude that the quake had devastated the church and caused its subsequent abandonment. Measuring about 85 feet long by 52 feet wide, the Aqaba basilica may be the oldest known example of a Christian church in the world (Parker 1998). In the absence of any inscriptions dating the construction of the church, Parker relied on unearthed artifacts such as pottery shards to estimate a construction date in the late third century (apparently even before the construction of Jerusalem's Church of the Holy Sepulchre). The destruction of the building is less of a mystery. Beneath a collapsed wall, Parker found more than a hundred coins minted during the reign of Constantius II, which lasted from 337 to 361. According to his hypothesis, the coins may have been part of an offering that fell from a collection box during the 363 earthquake, which caused massive destruction in the area.

Rome

Even when the fallen walls have been cleared away, the remains of great ancient edifices sometimes yield clues not only to the occurrence of earthquakes but also to special local hazards that have gone unrecognized even into modern times. The Roman Colosseum (or part of it) is an example of this. Most of the millions of people who visit this site do not realize why only part of the ancient arena remains standing, nor do they wonder why the parts still standing were spared. The entire southern half of the exterior wall is missing (Figure 4.14), having collapsed in an earthquake in AD 1349 that caused widespread damage in Rome, and even worse damage in the Alban Hills east of the city.

The strange way the elliptical outer "shell" of the Colosseum partially collapsed tells us that something unusual happens in that area during earthquakes. Without knowing more about the subsurface geology, it would be hard to draw any definite conclusions. In 1995, however, the geophysicist Peter Moczo and his colleagues conducted a seismic study of the ground beneath the Colosseum, using sound waves to make images of the subsurface structure (a sort of geological ultrasound technique). The images revealed that the damaged half of the Colosseum stands on relatively recent alluvium—accumulated river-borne sediment—which fills the prehistoric bed of an extinct tributary to the Tiber River. The other half of the structure stands on the older, more stable ground of the riverbank (Moczo et al. 1995).

A filled-in riverbed like this is a particularly insidious hazard during earthquakes. The flat, dry surface of such a buried river channel is an attractive place to build, offering easy digging of foundations, little need for leveling or grading, and an attractive, wide space for a grand edifice such as the Colosseum. Unfortunately, several factors make it particularly susceptible to earthquake shaking. Alluvial sediments are weak, since they are simply deposited by water and not cemented in place. In addition, although the ancient, buried river channel may have been invisible to the builders, its existence beneath the surface often results in a high water table in

Figure 4.14 The aerial view of the Roman Colosseum: the missing southern part of the exterior wall collapsed in an earthquake in 1349 AD.

the weak sediments. When weak, wet sediments are shaken in an earthquake, they tend to liquefy and flow out from under surface loads, such as heavy walls, effectively turning their foundations to a slurry.

Along with this inherent weakness of the foundations, the shape of the riverbed may amplify ground shaking. Earthquake waves are reflected off the harder sides of the buried river channel, in much the same way that echoes bounce off the walls of the Grand Canyon. These continuing "echoes" result in shaking that tends to be more severe and last longer in such geological settings.

The earthquake that destroyed the Colosseum must have been moderately large to cause the observed damage, but the southern side was especially badly shaken. Not only would the southern side experience stronger ground shaking, but because of complex interactions between the waves and the ground geometry, the southern half may have been shaking out of phase with the rest of the building, so that the waves might literally have ripped the southern half from the surviving structure. The earthquake was not so large as to shake down the more firmly founded northern half of the structure, leaving the remnants of its architecture for tourists to appreciate today. More than a historical marvel, though, the Colosseum can serve as a warning, alerting us to the danger of extreme shaking at this site and the need to plan for it in any modern construction along the hidden riverbed.

Jerusalem

Like Masada and the Colosseum, Jerusalem has experienced several recorded earthquakes and must have experienced many others in the more distant past. Unlike the previously mentioned sites, however, its earthquake-damaged walls were always either repaired or cleared away to make room for subsequent construction. Any evidence for past destruction that remains buried beneath the ground is usually inaccessible, since nearly every square foot of real estate in the city is either occupied, disputed, or both. Therefore, most evidence for earthquakes in this city is in the standing repaired walls, not in excavations of collapsed structures.

The image of a section of the wall of the Old City (see Figure 1.3) is a stunning example of the repair record: patches of different block sizes, stone types, and workmanship, ranging from the huge, well-worked blocks of Herod's affluent times (around 50 BC), to small stones, roughly worked, from when the city was poor. The latest level of repair dates to the latter part of the Ottoman Empire's rule. In the upper left corner of the photo, we can

Figure 4.15 A nineteenth-century drawing of the eastern wall of the first-century Judeo-Christian synagogue in Jerusalem, which became known as the Church of Apostles by the archaeologist Louis Hugues Vincent. The construction phases suggest a pattern of damage and repair. Some, perhaps most, of the patches in this image denote damage by several earthquakes during the past two thousand years.

see also the precursor for the next collapse—cracking caused most probably by the Jericho earthquake of 1927.

Buildings, too, show this record of continued repair from earthquakes. A sketch of the front wall of the so-called Judeo-Christian Synagogue in old Jerusalem, which dates back to the first century AD, is shown in Figure 4.15. The different styles of patches indicate at least six major repairs and other minor ones. Some of these repairs are from known conflicts, others are from known earthquakes, and the circumstances of still others are long forgotten. Certainly some are from earthquakes for which we have no written record.

DISPLACED ARCHES

Arches, as architectural elements, were introduced approximately twenty-five hundred years ago. They enabled ancient architects to create large, expansive rooms and wide halls using only small stone blocks. In earlier building schemes, the width of a room or hall was limited by the size of a single, horizontal stone roof slab or wooden beam, and by the weight that it could support. Massive slabs were necessary to support the weight of the masonry above even a relatively narrow door in a wall of stone. The horizontal stone slab was subjected to bending stresses, which cause tension in the lower half of the slab. Rocks fail much more easily under tension than compression, and thus this horizontal slab was the weak point in the construction.

In arches, however, weight is borne relatively evenly by each stone in the structure, with each stone in compression rather than tension. This new architectural method provided structures with great vertical strength and less weight, without the difficulty of obtaining and maneuvering the massive structural members necessary for the older style. The key to the strength of this structure, which could be built without mortar or pins, was that the weight of the stones compressed the arch, pushing each stone firmly against its neighbor. Unless a stone completely disintegrated under the pressure, a well-designed arch would stand indefinitely under its own weight. However, if a single stone were removed from the perimeter of the arch, the entire structure could collapse.

In ordinary circumstances, builders would not expect any of the stones to move, because the immense weight of a masonry structure holds them firmly in place. In the extraordinary circumstance of an earthquake, however, the earth's motion wields the structure's own weight against it, effectively throwing the whole edifice around. Moderate and large earthquakes can result in very high ground acceleration, occasionally even exceeding the acceleration of gravity, particularly in areas where the local topography or geology amplifies ground shaking (as under the south side of the Colosseum). An earthquake, therefore, can effectively counteract gravity, which obviously would have disastrous effects on structures that depend on gravity for their integrity. However, even when the motion is less severe, structures are subjected to both strong horizontal forces and vertical forces acting against gravity.

The key to the weakness of arch-based architecture is the ease with which the base of the arch can slide horizontally, especially if partially deprived of gravitational compression. Typically, even in structures built in the second half of the twentieth century, the bases of arches were not pinned in any way to the structures beneath them. Thus, the enormous horizontal forces generated in an earthquake can shift the base of the arch sideways on its foundation, relieving the compression holding the arch together, causing total collapse. Because there was no concept of seismic hazard in ancient times—no idea that earthquakes can be expected to repeat and no way to grasp the huge forces they generate—no thought was given to pinning the structures against horizontal motion as an anti-seismic precaution.

Most arches that were hit by sufficiently large earthquakes in the distant past simply collapsed, and all we find in the excavated rubble is a heap of stones that made up the arch. Unless the whole wall toppled neatly outward, like the example shown in Figure 4.16 from the ruins of Bet Shean's main street, the chaotic remains of collapsed arches are not very useful indicators of earthquake damage. They are too difficult to distinguish from collapse caused by slower earth movements or destruction by violent human activity. However, earthquakes sometimes stop short of destroying a building; if the base of the arch slips only a small amount, one or more of the wedge-shaped stones, called *voussoirs*, in an arch can drop down until it is again pressed tightly against its neighbors. Cases such as these, where the sliding of the arch's base was incomplete, are rare, but they can provide insight into the seismic history of a region.

Kala'at Namrud

The ruins of the huge Arab and Crusader fortification of Kala'at Namrud (alternately spelled Kalat Nimrod, or Kalat Namrood) is

Figure 4.16 This wall in the ruins of Bet Shean toppled outward into the street during the earthquake of AD 749.

situated on the main road from Galilee to Damascus. Among the collapsed towers, fractured walls, and huge stones scattered around the fortification are several arches partially intact. The most vulnerable aspect of this architecture is obvious: the voussoirs in the arches, pictured in Figure 4.17. On the left side of the photograph, indicated by arrows, we see that the entire base of the doorway slipped sideways, loosening the stones in the arch. Subsequently, the voussoirs in the arch dropped down, restoring the compression, and leaving the doorway a bit wider and shorter than the original architect had intended. Had the stones dropped another 10 centimeters, the doorway would have collapsed entirely, bringing the whole wall with it. The damage here was almost certainly from an earthquake, but there is no written record of when it occurred. A large earthquake that hit northern Israel, Lebanon, and Syria in AD 1202 is a good candidate, but we cannot be certain, because descriptions of that earthquake do not mention damage to Kala'at Namrud.

Figure 4.17 Slipped voussoirs in a doorway and vaulted passage at Kala'at Namrud, Israel. Scholars believe that the damage was caused by the earthquake of AD 1202.

Just like the individual arch in the doorway in Figure 4.17, the entire ceiling of the dark, vaulted passageway behind the doorway was endangered in the same earthquake. An entire row of stones in the arched ceiling slipped down, stopping just before total collapse. I can only imagine how frightening it might have been to be hiding in this passageway when the earthquake hit!

Rome

The half of the Colosseum that withstood the 1349 earthquake experienced a partial collapse similar to the slippage in the arch at Kala'at Namrud. The remaining structure, including the arches pictured in Figure 4.18, offers a glimpse of what happened to the rest. The arch that once stood to the left of these arches collapsed in the destructive earthquake, and although the leftmost arch shown in the photo did not fall, it only narrowly escaped. The movement was arrested just in time, as we can see in the close-up. The voussoirs slipped only partially, and the arch survived.

Baalbek

As shown in a hand-tinted lithograph by the nineteenth-century artist and explorer David Roberts (Figure 4.19), a massive keystone slipped dramatically in the entrance to the temple of Bacchus at Baalbek in northeastern Lebanon, built by the second-century AD Roman emperor Antoninus Pius. As Roberts (2000) wrote in his journal, "Earthquakes have shaken this extraordinary remnant; but from the magnitude of the blocks which form the lintel, the central one, being wedge-shaped, has slipped only so far as to break away a portion of the blocks on either side, and thus remain suspended."

A well-documented earthquake that struck the Bekaa Valley in 1759 is known to have damaged Baalbek, but at least nine other quakes with estimated magnitudes greater than 6.5 have been mentioned in historical accounts since AD 1100. This region is shaken by earthquakes on two fault systems: the northernmost section of the Dead Sea fault system and the very active East Anatolian fault system. The seismic hazard for other cities in the region remains poorly understood.

California

The same effect on arches is also found in modern buildings, such as the arched entrance to the Long Beach High School right after

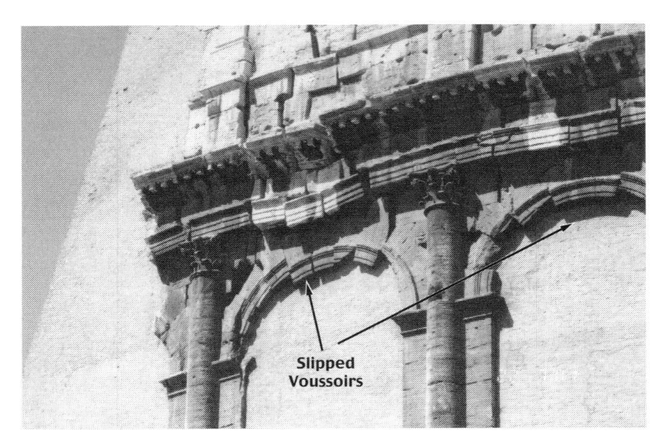

Figure 4.18 Close-up of the remains of the outer Colosseum wall, preserved in its state of disarray. The left-most arch came very close to collapsing.

Figure 4.19 A massive keystone slipped dramatically in the entrance of the temple of Bacchus at Baalbek, in northeastern Lebanon.

Figure 4.20 Slipped keystone at Long Beach High School, California, 1933 (from Wood, H. O. Preliminary report on the Long Beach earthquake. *Bulletin of the Seismological Society of America* 23, 2, 43–56, 1933, © Seismological Society of America).

the 1933 Long Beach, California, earthquake (Figure 4.20). This earthquake's magnitude was a fairly moderate 6.3, so slippage of the base was not large enough to topple the entire entrance.

The larger 1906 San Francisco earthquake (magnitude 8.2) caused more than $2 million worth of damage at the Spanish-Mission-style campus of Stanford University alone. The heavy damage included, ironically, the geology corner of the main quadrangle, which was surrounded by an arched colonnade. Figure 4.21 is a 1906 photo of one of its arches, almost collapsed and still propped up by wooden bracing. Again, the displacement at the base of the column that supported the arch caused the arch to widen and the keystone to drop almost completely through, so that the arch barely survived. It was repaired after the 1906 earthquake, but the base was not pinned in that repair, as if the vulnerability of the arch construction were still not fully appreciated. Thus, eighty-three years later, the same arches slipped again in the smaller 1989 Loma Prieta earthquake (Figure 4.22). Afterward, all the hundred or so columns in

Figure 4.21 Damaged arch at Stanford University, California, 1906 (Courtesy Stanford University Archives, SC034 Branner Papers).

Figure 4.22 Arch damaged again at Stanford during the 1989 Loma Prieta earthquake (Linda A. Cicero/Stanford News Service).

the Stanford main quadrangle were finally pinned to their bases. This measure should help reduce future damage.

MONUMENTS

Besides the damage to the unreinforced-masonry buildings, several small monuments around the Stanford campus were toppled from their bases during the 1906 earthquake. Figure 4.23, one of my favorite photographs, was taken at Stanford right after the disaster; the gravity of the statue's pose and the eminence of the name carved on the base, "L. Agassiz," stand in wonderful contrast to the foolishness of perching a stone statue high on a narrow base in earthquake country. The university officials who held Louis Agassiz in such high regard had their heads if not in the pavement, at

Figure 4.23 A statue of the glacial geologist Louis Agassiz toppled from its perch on the roof of a Stanford University building, some five miles away from the ruptured San Andreas Fault during the 1906 earthquake (Courtesy United States Geological Survey).

least in the sand, unaware, as they were, of the seismic hazards directly under their feet. Still, out of this destruction was born a burst of scientific research and innovation in the areas of earthquake engineering, seismology, and geology. Today the retrofitting of older structures at the university is an ongoing project in the effort to avoid a repeat of the 1906 and 1989 losses.

Monument building is a long-standing human obsession. We build them high into the sky, as if height were somehow correlated with the greatness of the person, city, or event we wish to commemorate. From the Statue of Liberty in New York to the Eiffel Tower in Paris, these monuments serve no practical purpose but to impress with their height and grandeur.

Ancient monuments were no different. Whether obelisks, towers, or statues, they were built to loom over their surroundings and dwarf onlookers, often straining the physical limits of the materials and methods used to construct them. This has made most monuments extremely vulnerable to earthquakes.

Rhodes

The Colossus of Rhodes in Greece, counted as one of the Seven Wonders of the ancient world, was a statue of the sun god Helios and was approximately 110 feet high. It was erected by the sculptor Chares of Lindos around 300 BC beside the port of Rhodes (Figure 4.24). The statue, which inspired the Statue of Liberty in New York Harbor, stood for only around sixty years before an earthquake destroyed it.

This earthquake struck the island of Rhodes in the middle of the third century, but the exact date is unknown. As noted by Guidoboni et al. (1994), contemporary accounts used as references by modern writers make the probable date between 234 and 218 BC, but modern earthquake catalogues disagree on the date. Further investigations of the geology between the modern towns of Rhodes and Kalitea in northern Rhodes show that a large earthquake hit

Figure 4.24 A drawing of the Colossus of Rhodes appeared in the 1612 book "Minerva Britanna" by Henry Peacham. The huge statue, one of the Seven Wonders of the Ancient World, was devastated by a massive earthquake in the mid-third century BC.

this area at around the appropriate time. The evidence includes an ancient wave-cut terrace on the limestone cliffs eleven feet above current sea level, traced back to the third century BC. This means that an earthquake (or earthquakes) around the third century lifted these cliffs eleven feet above their previous height (Higgins and Higgins 1996). In any case, the earliest literary sources, including Strabo, Polybius, Diodorus, Pausanias, and Pliny, often mention the event. Pliny, in *Natural History* (1938 [ca. AD 77], 34 (18): 41–42), describes the ruined statue, which remained for almost a thousand years where it had fallen beside the port: "Sixty-six years after its erection the statue fell over in an earthquake, but even lying down it is a marvel . . . where the limbs have been broken off, there are huge gaping cavities. Inside it, one can see rocks of enormous size which Chares had used to stabilize it when he was building it."

This quake, whose magnitude has been estimated at 7.5 (Papazachos et al. 2000), also destroyed other regions of the Aegean, including Lycia and Caria, as recorded by Pausanias (1918, 2.7.1):

> When they had lost their power there came upon them an earthquake, which almost depopulated their city and took from them many of their famous sights. It damaged also the cities of Caria and Lycia, and the island of Rhodes was very violently shaken, so that it was thought that the Sibyl had had her utterance about Rhodes fulfilled.

The original *Sibylline Oracles* were burned partially in the last century BC and completely in the fourth or early fifth century AD. However, copies of the manuscripts believed to date back to a period between 100 and 600 AD include a passage that may be the one to which Pausanias refers:

> When, earth being shaken by earthquakes, cities fall.
> And sand shall hide all Samos under banks.
> And Delos visible no more, but things
> Of Delos shall all be invisible.
> And to Rhodes shall come evil last, but greatest.
> (*Sibylline Oracles,* bk. IV, 125–129)

The people of Rhodes declined to rebuild their statue, thinking that perhaps they had offended Helios with their arrogance and the earthquake was their punishment. In the seventh century AD, the last iron and bronze remains of the colossus were finally broken up for scrap metal.

Rome

Rome, of course, is a city of monuments, many of them affected by earthquakes in this seismically active region. Some monuments were completely destroyed and others have been repaired, but one that preserves its earthquake damage is the Column of Marcus Aurelius (Figure 4.25a), erected in the second century AD, along with the column of Trajan, to celebrate military victories.

Figure 4.25 The Column of Marcus Aurelius (a) still stands in Rome, despite earthquake damage. The close-up of the column (b) shows rotation between two of its massive stone elements. (from E. Boschi et al. Resonance of subsurface sediments: An unforeseen complication for designers of Roman columns. *Bulletin of the Seismological Society of America* 85 (1) 320–324, 1995, © Seismological Society of America).

Renato Funiciello, a leading professor of geology in Italy, arranged to take me to the top of the Column of Marcus Aurelius, situated in the Piazza Colonna in the heart of the classical city. The column stands directly across from the office of the Italian prime minister. By climbing a helical staircase inside the column, one emerges on the top of the monument, where one can see directly into the Italian prime minister's inner office and, if he is sitting at his desk, smile and wave at him. Obviously, the security ramifications of this situation made the climb a complicated matter, even in the late 1990s, when the fear of terrorists was less intense than it is today. However, after arranging matters with the Roman Police and the National Assembly Police, and after securing help from some well-connected friends, Funiciello was finally able to take me to the top of the monument.

The purpose for our climb was to examine from the inside the damage the column had sustained in an earthquake some time after its construction. The column, approximately 3.7 meters in diameter and 42 meters high, is constructed of seventeen round blocks of Carrara marble, each weighing 30 tons, stacked on top of one another. Early sixteenth-century prints show that the column once sported large vertical fractures, which Pope Sixtus V ordered repaired during the latter part of that century. However, still visible in the elaborate carvings on the surface of the column is an 8 cm offset between two of the massive blocks, including a small component of rotation (Figure 4.25b), between two of the massive blocks. Although no historical details record the cause of this damage, we suspect it occurred during one of the many earthquakes that shook the area during the first fourteen centuries of the monument's existence.

Funiciello, in collaboration with other scholars (Boschi et al. 1995), studied how resonance of subsurface sediments in earthquakes may have affected the Marcus Aurelius column differently from the Trajan column, which still stands 700 meters away. Although the two columns are of similar construction, the former was significantly damaged by past earthquakes whereas the latter apparently survived unscathed. Like the Colosseum, these columns represent another case where archaeological evidence of earthquake damage has led to a better understanding of the geology beneath a modern city, which may help builders take measures to alleviate seismic hazards in future construction.

Because architectural and historical treasures are so densely distributed in the city of Rome, and because Rome is in a zone of significant seismic hazard, Roman archaeologists and engineers spend a great deal of energy and research money trying to understand how earthquakes affect ancient buildings. The architectural relics of the ancient Roman Empire are a major attraction in Italy, and a significant earthquake there, besides destroying some of these treasures, could kill countless tourists and scholars. A 1997 earthquake in Assisi, Italy, which devastated the Cathedral of St. Francis of Assisi, brought the reality of the danger into sharp relief: as two Franciscan friars and two surveyors from the Culture Ministry were documenting the damage the quake had inflicted on frescoes inside, an aftershock toppled the building, killing them all.

Luxor

After giving a lecture on archaeology and earthquakes at a meeting in Cairo, Egypt, I happened to fall into a conversation with a professor from Cairo University, who then took me to Luxor to see the ruins of the Ramesseum, the burial temple of Ramses II, arguably the greatest pharaoh of Egypt. Ramses II died in 1212 BC, after governing Egypt for sixty-six years.

The huge monument erected to commemorate Ramses II is impressive: a sixty-foot statue of the pharaoh sitting on his throne, carved from a single piece of granite weighing close to a thousand tons. This enormous chunk of granite was floated down the Nile from a quarry in Upper Egypt. The effort and planning required for this feat were amazing, and, to this day, no one knows for certain how it was done. Unfortunately, the monument now lies in pieces among the ruins of the Ramesseum (Figure 4.26). In fact, the remains of this colossal monument gave rise (through a complicated route involving mistranslation by an ancient scholar) to Percy Bysshe Shelley's famous poem, "Ozymandias of Egypt":

> I met a traveler from an antique land
> Who said: Two vast and trunkless legs of stone
> Stand in the desert. Near them on the sand,
> Half sunk, a shattered visage lies, whose frown
> And wrinkled lip and sneer of cold command
> Tell that its sculptor well those passions read
> Which yet survive, stamped on these lifeless things,
> The hand that mocked them and the heart that fed;
> And on the pedestal these words appear:
> "My name is Ozymandias, king of kings:
> Look on my works, ye Mighty, and despair!"
> Nothing beside remains. Round the decay
> Of that colossal wreck, boundless and bare,
> The lone and level sands stretch far away.

My trip to Luxor was in the off-season for tourists, so the site was not particularly crowded. Nevertheless, my geophysicist's approach

Figure 4.26 The huge monument erected to commemorate Ramses II and destroyed by the 27 BC earthquake now lies in pieces among the ruins of the Ramesseum in Egypt.

to the ruins must have seemed odd to the few off-season tourists and local boys who were there; as they gathered around the impressive monuments and columns still standing, I, with my back to the glory, was far more interested in what had fallen. The columns of the Ramesseum lie partially collapsed, and the outside of its great stone façade—often called the first pylon—is in ruins. I believe that much of this damage, as well as damage in Karnak and Luxor, just across the Nile to the east, was caused by an earthquake in 27 BC. This earthquake was probably also responsible for damage to the two huge statues of Amenhotep III, creating a fissure in one of them that would "sing" when warmed by the morning sun. The Greeks thus knew the statues as the Colossi of Memnon, after the son of Eos, the goddess of the dawn. Unfortunately, the Roman emperor Septimius Severus tried to have the statues repaired, and, in doing so, he silenced them.

The glory of its aboveground monuments notwithstanding, the Luxor area is most famous for its splendid tombs. The nearby Valley of the Kings experienced very little damage from the 27 BC earthquake, an observation scholars have used to argue against earthquake damage as an explanation for aboveground destruction. The fact is, however, that buried tombs and small caves are remarkably earthquake-resistant structures, as long as rooms are small and have no wide, unsupported ceilings. Egyptian tombs, with their small chambers and narrow connecting shafts, are quite impervious to all but the most severe temblors.

The 27 BC earthquake raises some important questions. Where was the epicenter of the earthquake that caused this destruction? From our current knowledge of the pattern of active faults, earthquake epicenters, and plate boundaries in this region, we have not been able to identify the fault responsible for this earthquake. In fact, were it not for our knowledge of that destructive event, we would not consider this an area of major seismic hazards. Since we know, however, that the 27 BC earthquake occurred, we must consider the possibility that the Luxor area is susceptible to large, mid-continental earthquakes occurring far from plate boundaries.

Although this type of earthquake is uncommon, Luxor is not the only place subject to this kind of risk. In 1811 and 1812, a series of massive earthquakes occurred in the United States, near New Madrid, Missouri, far from any plate boundary. We still do not fully understand the stresses that caused these earthquakes, but the New Madrid earthquakes were among the strongest known to occur in the continental United States, and they indicate an enormous seismic hazard for the lower Midwest. Investigations in the New Madrid area show that quakes there have a very long recurrence interval, with estimates varying from two hundred to more than ten thousand years of quiescence between events. If the new Madrid earthquake had occurred a few hundred years earlier, before European settlers and written records, we would be unaware of any seismic hazard there. Archaeological evidence of these large, intra-continental earthquakes elsewhere in the world may be the only indication we have of increased seismic risks in regions where the historical record is not sufficiently long to have captured such rare events.

Figure 4.27 An old Italian print depicting the Lighthouse at Alexandria.

Alexandria

Another wonder of the ancient world, the Lighthouse, or Pharos, at Alexandria, Egypt (Figure 4.27), is also thought to have collapsed in an earthquake. As nearly as archaeologists and historians can determine, the structure, first built around 290 BC by Ptolemy Soter I and his son, was damaged by several earthquakes during the many centuries of its existence. The exact details, however, remain controversial.

An expedition led by the French archaeologist Jean-Yves Empereur, which commenced in 1994 on the island of Pharos off the coast of Alexandria, discovered the remains of the great ancient lighthouse. Ancient and modern texts describe the monument as around 380 feet high.

John and Elizabeth Romer (1995, 55) attributed the structure's final destruction to an earthquake in AD 1375, citing unspecified Egyptian documents as their source. Ambraseys, Melville, and Adams (1994) proposed that a quake in AD 796, in the month of April, cost the lighthouse its upper level, and scholars agree (e.g., Ambraseys, Melville, and Adams 1994; Guidoboni and Comastri

1997; and Sieberg 1932) that the historical earthquake of August 8, 1303, shortened it further. However, the argument continues about the dates of all the Alexandria earthquakes, as well as the date of the structure's final collapse. We know it finally tumbled into the sea sometime in the fourteenth century, possibly as a direct result of the August 8, 1303, earthquake. This massive seismic event in the Mediterranean was felt in a huge area, including Crete, Egypt, Rhodes, Jordan, Syria, Palestine, Turkey, and Cyprus, and it was followed by a large tsunami (Sieberg 1932; Guidoboni and Comastri 1997). Damage was particularly heavy in Crete and Alexandria. As with other geographical areas covered in this book, the main point here is not to become embroiled in debates over the precise timing but to uncover evidence of important destruction by earthquakes in antiquity, and thereby estimate the very long-term seismic hazards of ancient sites.

TSUNAMIS, SEICHES, AND SEISMITES

The December 26, 2004, Sumatran earthquake, and the devastation it visited on the beaches and coastal towns of the entire Indian Ocean region, is a stark example of the destructive power of tsunamis. Wherever earthquakes move the sea floor, there is the potential for these destructive seismic sea waves.

As I mentioned in the previous chapter, the island nation of Greece has always been particularly susceptible to tsunamis, and, in fact, the earliest reference to a tsunami comes from Thucydides (1910), who described the inundation of several Greek towns following an earthquake during the Peloponnesian War, in 426 BC:

> The next summer the Peloponnesians and their allies set out to invade Attica . . . but numerous earthquakes occurring, turned back again without the invasion taking place. About the same time that these earthquakes were so common, the sea at Orobiae, in Euboea, retiring from the then line of coast, returned in a huge wave and invaded a great part of the town, and retreated leaving some of it still under water; so that

> what was once land is now sea; such of the inhabitants perishing as could not run up to the higher ground in time . . . [List of destroyed ports deleted] . . . The cause, in my opinion, of this phenomenon must be sought in the earthquake. At the point where its shock has been the most violent the sea is driven back, and suddenly recoiling with redoubled force, causes the inundation. Without an earthquake, I do not see how such an accident could happen.

Thucydides was right. Tsunamis occur when the sea floor moves either up or down, usually in either a normal- or reverse-faulting earthquake, although small tsunamis have been associated with undersea landslides as well. The sudden displacement of the sea floor pulls a large volume of water along with it, briefly shifting the sea level in a broad region above the sea-floor disturbance. Of course, water cannot remain in a heap, and so it spreads out from the area of disturbance in a wave, as if a giant stone had been dropped in the ocean. These enormously broad waves can travel for thousands of kilometers.

On the open ocean a tsunami is barely noticeable, because it is so very broad (typically covering hundreds of square kilometers) and flat (only a few meters high, at the most). Such a broad, flat swell would be unnoticeable amid the general chop at sea. However, when the wave approaches the shallower water of a coastline, it transforms; the water piles up until, like a smaller ocean wave that laps the shore under ordinary circumstances, it breaks, with devastating consequences. Seismic sea waves have been known to run up onto coastlines to heights of as much as 40 meters.

Often tsunamis are preceded by a sudden withdrawal of water from the shore. This can cause serious hazards, because onlookers, curious to examine the now-exposed sea floor, are lured down onto the beach, only to be swept away by the returning water. Only by publicizing this particular earthquake hazard can science help to alleviate this risk.

Oceans are not the only waters subject to earthquake-generated waves. Smaller waves can be generated in lakes and inland seas, even by distant earthquakes. Such waves are known as *seiches*. Seiches measured in 1955 in Norway and England were generated by an earthquake in Assam, India. Seiches were frequently reported on both the Dead Sea and the Sea of Galilee during Mediterranean-area earthquakes in antiquity. The earthquakes of AD 363, 749, and 1546, which are described in later chapters, all caused seiches in the Dead Sea.

When earthquakes shake the floor of an ocean or lake severely enough, they can cause the sediments to liquefy, in much the same way as I described earlier in relation to ground disturbances during earthquakes. However, because the entire sea floor or lakebed is covered with water-saturated, loose sediments, the liquefaction can be extremely widespread and lead to underwater landslides, which in turn can cause tsunamis and seiches. The other side effect of these landslides, and a fortunate one for archaeoseismology, is that a record of the earthquake is preserved in the chaotically mixed sediments that flowed during the shaking. These characteristic mixed layers found in sea-floor sediments are called *seismites*. Often, the mixed layers, and the undisturbed layers deposited over them, contain organic material that can be used to establish radiocarbon dates for the earthquakes. In chapter 7, I discuss how these seismites were used in one Dead Sea study to confirm many historical accounts of earthquakes in the Middle East.

FIRES

From water, we turn to fire. Certainly the most equivocal form of earthquake damage that can be preserved in the ruins of ancient cities is widespread fire damage. Ash layers are extremely common in archaeological digs, and they often can be used as dividing lines between one layer of habitation and the next.

The problem is that fires can arise from any number of causes, and as modern arson investigations show, it can be next to impossible to determine the cause after the fact. Therefore, in this section, instead of describing the remnants of earthquake fires from a series of archaeological sites, I will explain how conflagrations

Radiocarbon Dating

The dream of archaeologists, paleontologists, and anyone else interested in the past is to have some test that can be used to determine the precise chronological age of any item. No test is available that can do that unambiguously for all objects. The closest we have, radiocarbon dating, or ^{14}C dating, has many limitations and complications.

Radiocarbon dating relies on the fact that carbon, the major building block of life, comes in several isotopes. The nucleus of a carbon atom always has six protons, but can have six, seven, or eight neutrons. These different varieties of carbon, called isotopes, derive their names from the sum of their protons and neutrons: ^{12}C, ^{13}C, and ^{14}C. Of carbon's three isotopes, the heaviest, ^{14}C, is radioactive. Formed constantly by neutron bombardment of nitrogen in the atmosphere, it is unstable and breaks back down into nitrogen over time.

Any living organism consumes carbon from the environment, so the ratio of ^{14}C to ^{12}C in its tissues remains approximately the same as that in the environment while it is alive. When an organism dies, however, the ^{14}C decays, and no more is added. Thus, we can determine the amount of time that has elapsed since the organism's death by measuring the ratio between ^{14}C and ^{12}C in the organism's tissues. In archaeology, this means, theoretically, that we can measure the age of wood, plant and animal fibers, and human remains.

In practice, however, there are many complications. Most fundamental is that there is no way to measure how much radioactive carbon was present in the original organism. If the ^{14}C decayed to something exotic that was not usually present in living organisms, then we could use the ratio of the ^{14}C and the decay product (called a "daughter isotope") to find the age. However, ^{14}C decays to ^{14}N, nitrogen's most common isotope, which is present abundantly in all living things. When radiocarbon dating was first developed, scientists assumed that the ratio of ^{14}C to ^{12}C in the atmosphere was constant through time. However, comparison of radiocarbon dates and tree-ring dates for very ancient trees has shown this to be false. Thus, original ^{14}C ratios must be determined from "wiggle traces" empirically determined from tree-ring data. The older the sample that is examined, the less accurate are the wiggle traces for that period, and so accurately dating very old materials is difficult. Radiocarbon dates for events in the Bronze Age are generally thought to be several hundred years younger than actual calendar dates.

There is also a limit on the range of dates the method can cover; ages younger than one hundred years and older than about fifty thousand years are difficult to determine, since, in the first case, there has not been time for measurable decay, and, in the second, nearly all the ^{14}C has been depleted. Another problem is that what happens to the organism after it dies can alter the amount of ^{14}C in its tissues. Groundwater infiltration or contamination from younger material can artificially "reset" the clock, making a date completely meaningless. It is sometimes difficult to determine whether this has occurred.

Another difficulty is that the ratio of ^{14}C can vary regionally, usually because of long-term input of CO^2 from a non-atmospheric source. For example, the ^{14}C in deep earth sources has long since decayed away, so a significant volcanic CO^2 source may throw off the dating in the region nearby. Regional corrections are far scarcer than global wiggle traces.

Finally, as in any other measurement technique, radiocarbon dating is subject to uncertainty; all we can determine is a probability that an object's age falls within a given range. With all these caveats, scientists must handle radiocarbon dating very carefully. It is important to recognize, too, that although dates are given as so many years "BP," "before present," they are, according to convention, actually "before 1950."

have arisen as a direct consequence of earthquakes, probably for as long as humans have wielded their unreliable mastery of fire. I belabor the point, because historians or archaeologists sometimes use evidence of widespread fire in a destruction layer to argue against an earthquake as the destructive agent.

Fires are ubiquitous following modern earthquakes. This is largely because of the predominance of wood in modern buildings, as well as the near universality of gas and electrical supply lines, both of which are vulnerable to earthquake damage. Thus, earthquakes in today's cities often ignite hundreds of small fires in the rubble, fires that are hard to extinguish because of blocked streets and overtaxed emergency-response teams.

Several assumptions have led many people to believe that conflagrations following earthquakes are solely a modern phenomenon. One example is the following passage by Robert Drews (1993, 39):

> Earthquakes, on the other hand, were in antiquity not associated with devastating fires, presumably because there were no gas mains or electrical cables, and most cities and towns consisted primarily of masonry structures. In his discussion of earthquakes Pliny does not even mention the danger of fire. . . . Of the several hundred ancient earthquakes that W. Capelle cataloged from literary sources, none is known to have ignited a city-wide fire. It therefore strains credulity to suppose that a single earthquake should have resulted in conflagrations at three sites in the Argolid—Mycenae, Tiryns, and Midea—and that similar fires should have been set by this or other quakes at Knossos, Troy, and Ugarit.

Drews has made two assumptions: first, that because ancient builders mainly used stone and mud brick, there was little to burn; and, second, that in the absence of modern utilities, there is little to cause a fire. Yet, in the same argument, he states that sackers routinely burned cities after looting them. Why are there enough combustibles to burn when kindled by looters but not by accidental causes?

The assumption that ancient sites had few combustibles stems partly from an unavoidable bias in archaeology: noncombustible materials such as stone, mud brick, and metal are more readily preserved than combustible organics such as wood, fabric, and straw. Just because these materials are rare in archaeological finds, however, it is not therefore reasonable to assume they were absent in ancient life, any more than they are absent in regions today where ancient building practices persist. The very commonness of ash layers in ancient archaeology argues against this, unless every ravaging army carried its own fuel to spread evenly over every city it sacked.

We also know that the second part of Drews's argument does not hold up, as there are many examples of fires caused by earthquakes and other accidents in cities without modern utilities. In Tokyo, in 1927, most of the fires that led to the city's near total destruction came not from modern utilities but from charcoal fires that ignited straw mats in houses throughout the city. In Lisbon, Portugal, after a devastating earthquake in 1755 (described in detail in chapter 9), cooking fires ignited a citywide conflagration that burned for three days, causing more damage (though not more deaths) than the earthquake shaking itself. In 1356, an earthquake in Basel, Switzerland, triggered weeklong fires that destroyed the city. Most accounts of the AD 363 earthquake in Jerusalem describe fires that killed many people. Further, the destruction from the earthquake of AD 17, which encompassed twelve cities along the Gediz River in Turkey and was described by Pliny (1938) as the worst quake in the memory of mankind, was aggravated by the accompanying conflagration (Ambraseys 1971). Clearly, we know that fires were associated with ancient earthquakes, though sources that never mention them abound.

Fires were not as ubiquitous in antiquity as they are in modern times. In modern cities, after all, electricity and gas lines are always on, regardless of the season or time of day. In antiquity, open flames were the source of ignition, and were more common around mealtimes (when food was being cooked), after dark (when lamps were used), and in colder seasons where fires would be used for heat. Thus earthquakes that occurred in warm seasons, or during times of day when fires were not stoked, would be expected to cause fewer fires.

Certainly, fires cannot be used to either rule out or implicate earthquakes. Rather, an ash layer indicates only that a city was burned, not why or how it was burned. In cases where other evidence strongly indicates an earthquake, however, evidence of fire may help provide a fuller picture of the damage the earthquake caused. As I show in the next chapter, fire can even help preserve details of earthquake destruction by baking mud bricks and tablets into a durable state, and even scorching the bones of earthquake victims.

CHAPTER 5

Under the Rubble

HUMAN CASUALTIES OF EARTHQUAKES

A great earthquake suddenly changes all the ways of life.
—Charles Davison, 1931, referring to the great Kanto earthquake in Japan, 1923

Although it is morbid to devote an entire chapter to describing the crushed remains of one earthquake victim after another, the grim recitation that follows serves an important purpose. As noted earlier, many archaeologists are reluctant to invoke earthquakes to explain excavated destruction. Instead, they attribute collapsed walls, uniformly fallen columns, and slipped keystones to poor construction, soil creep, and ground water seepage, if not just aging in general. Invaders are often blamed for destroyed monuments and fires. Crushed and broken skeletons found under rubble are a different story, however. Many archaeologists accept these as definitive evidence of earthquake destruction.

The destruction of a massive stone building by human hand takes some time. Given the methods employed in ancient times, people had ample time to escape when an army was hammering at a building; those inside were likely to flee the structure rather than wait for the roof and walls to crash down upon them. An earthquake, on the other hand, gives no warning, and often there is no time to do anything but cower until the shaking stops. Thus, the discovery of human remains crushed beneath their dwellings creates a strong argument for an earthquake, rather than a militia, as the destructive force.

Lack of skeletons, however, is not always an argument against earthquakes, because not all destructive earthquakes kill a large number of people. The time of day or season when an earthquake strikes has a lot to do with the number of casualties it causes. We see this in modern quakes, as well as ancient ones. Some of the worst disasters in human history were earthquakes that occurred at night, when people were asleep and unable to react quickly to the shaking. The 1976 earthquake in Tangshan, China, occurred in the early morning, at 3:42 AM, while the city was still asleep, and it was the worst earthquake disaster of the twentieth century, killing from 250,000 to 600,000 people, perhaps even more. Chinese government estimates for deaths in this earthquake are acknowledged to be, for political reasons, unreasonably low.

Another early morning earthquake struck Killari, India, at 3:56 AM, on September 30, 1993, registering a magnitude between 6.2 and 6.4 and affecting the regions of Latur and Osmanabad, in India's Maharashtra state. Considered the deadliest earthquake of 1993, it left more than 9,000 people dead and about 30,000 wounded. In the village of Killari, about 30–35 percent of the estimated 15,500 residents lost their lives. Most were found buried under the rubble of their rural mud-and-brick or stone residences. Nearly all the buildings in Killari were destroyed; survivors were mostly those who had lived in wood-frame houses (Narula 1995).

Conversely, the much larger Bihar-Nepal earthquake (magnitude 8.4) of July 15, 1934, struck in the afternoon, when most residents were outside their homes. Although the death toll was still large—7,253 killed in India, and 3,400 in Nepal (Richter, 1958)—the number of casualties was considered low for such a large earthquake; at two orders of magnitude larger than the Killari earthquake, the shaking would have been a hundred times as severe, and nine hundred times more energy would have been released. It is difficult to imagine the number of fatalities that would have occurred had the earthquake struck as the city slept.

The skeletons that were found crushed under stones in Shanidar Cave in Iraqi Kurdistan, as described in chapter 2, represent a unique situation; there is really no explanation besides earthquakes for the layers of collapse that the excavator Ralph Solecki uncovered. In no way could an enemy have caused the cave to collapse. In most excavations, however, there is always room for argument. Skeletons found under rubble may be evidence of an earthquake, but one might also argue that the person was killed by human hands and that the skeleton was buried by later collapse unrelated to the cause of death. In these cases, we often turn to forensic anthropology, a field most commonly devoted to solving modern-day crimes, to extract more information from the bones of ancient victims. Sometimes we can determine more about the circumstances leading up to the death and definitively describe the types of injury that were inflicted on the person in question. In other instances, evidence clearly shows that the victim was alive, or at least only recently dead, when buried by debris.

ANEMOSPILIA, 1700 BC

One of the strangest examples of crushed skeletons was uncovered on Crete by a 1979 expedition from the University of Athens. At Anemospilia, on a hillside near Arkhanes—a modern village four miles from the site of the ancient Minoan palace of Knossos—Yannis Sakellarakis and Efi Sapouna-Sakellaraki discovered the ruins of a Minoan temple. Early in their excavation, when they had exposed three of the four rooms in the temple, they suspected that it had been destroyed around 1700 BC by one of a sequence of large earthquakes that had also destroyed many palaces on Crete. They based their conclusion on pottery, artifacts, and other archeological evidence found at the site. One skeleton was discovered crushed and deteriorated to the point where its sex could not be determined (Sakellarakis and Sapouna-Sakellaraki 1981, 222). After the collapse, any combustibles in the temple had burned, leaving a layer of ash and scorch marks on the stones. Only when the excavators cleared the last room, however, did the full drama of the destruction unfold.

The fourth room of the temple yielded the skeletons of two more temple functionaries, presumed to be a priest and his female acolyte. The skeleton of the priest wore a ring of silver and iron (a precious metal in the Bronze Age) and an exquisitely engraved seal tied on his wrist, indicating that he was a person of power and high rank. The priest and his female attendant lay before an altar of the sort that often decorated Minoan frescoes. The altars depicted in the art of the time were used for blood sacrifices, usually of bulls, to the Minoan gods. Indeed, the earthquake seemed to have interrupted a religious rite in full swing.

This sacrifice was exceptional, however. The bones found on this altar were not those of a bull but of a young man about eighteen years old and in apparently perfect health, reclining on his right side, and, from his position, probably bound hand and foot (Figure 5.1). Lying on his skeleton was a ceremonial bronze knife sixteen inches long and weighing more than a pound, with an ornately incised blade—the type of knife thought to have been used for animal sacrifices (Sakellarakis and Sapouna-Sakellaraki 1981, 220, 222). Had the young man died in the collapse, before the priest could take his life? The excavators turned to the medical anthropologist Alexandros Contopoulos of the Athens Medical School for more clues to his death. As Dr. Contopoulos explained,

> When a body with its blood supply intact is burned, the bones turn black. But if the blood has been drained before the fire, the bones will remain white. When we looked closely at this skeleton, we saw that the bones of the left side, which was uppermost, were white, while those on the right side were black. Thus, I believe that half of this man's blood had been drained before the fire. The loss was more than enough to kill him. The heart stopped pumping, leaving blood still in the lower side. (Sakellarakis and Sapouna-Sakellaraki 1981, 218)

The unlucky temple functionaries had been caught in the act of a rare human sacrifice. Was this sacrifice their response to earth-

Figure 5.1 Destruction at Anemospilia. An eighteen-year-old man, whose skeletal remains were unearthed by archaeologists, is believed to have been a sacrificial victim. The ceremony was interrupted by an earthquake that struck the temple (Martha Cooper/National Geographic Image Collection).

quake foreshocks, an attempt to appease the gods and stave off a major disaster, as Sakellarakis and Sapouna-Sakellaraki (1981) and Alsop (1981) suggest? If so, clearly it did not work. We can only speculate, as no written account of this disaster has been found.

TEL DOR, 1100 BC

The destruction of the city of Dor is another disaster for which we have no historical account, and for which the forensic examination of a skeleton has become a key piece of evidence. Located in Israel on the Mediterranean, just south of Mount Carmel, this port city was occupied over the centuries by Canaanites, Sikils, Phoenicians, Israelites, Assyrians, Babylonians, and Persians. It was no stranger to armed conflict. Ephraim Stern (1993), who excavated here for several seasons, discovered that one habitation layer of Dor was burned and destroyed around 1050 BC, with a widespread layer of debris and ashes over the entire site:

> From this period [ca. 1050 BCE], we found massive evidence of a fierce conflagration that had oxidized the mud bricks and shattered the limestone used in the buildings, leaving great areas of ash and charcoal as much as 6 feet thick. . . . The same thick destruction layer resulting from a violent conflagration appeared on the other (western) side of the mound.

This find raises an obvious question: Who, or what, was responsible for this conflagration? Stern believes that "Dor was attacked and destroyed by the Phoenicians in the course of a struggle for control of marine trade routes, as undoubtedly happened also at the Sea People city of Acre and at other coastal cities held by the tribe of Sherden." In the absence of a better explanation, this is a reasonable scenario. However, on the last day of the 1992 field season, the excavators at Dor made a startling discovery that was to invite reinterpretation of the entire sequence of events.

As the excavators were cleaning up and preparing for the last round of photos before closing the site for the season, one group suddenly uncovered the bones of a human foot, protruding from beneath a blanket of rubble. With the whole crew working feverishly in shifts for the rest of the day, an entire skeleton was gradually revealed (Figure 5.2). Andrew Stewart (1993) of the University of California, Berkeley, describes the skeleton as he and his crew uncovered it:

Figure 5.2 "Doreen," a probable earthquake victim at Dor in Israel, was crushed by a fallen wall (after Andrew Stewart/*Biblical Archaeology Review*, 1993).

> This was by no means easy archaeology. The room was small, and made smaller by the low stone screen against which she lay. . . . She was both badly contorted and cruelly smashed up. A limestone wall had fallen on her and had crushed her into the earthen floor below. Numerous rocks had penetrated the skeleton itself. A scatter of potsherds, stone tools, a bone needle and several small animal bones lay right beneath her, some of them also poking into her body.

Stewart, hailing from earthquake country himself, immediately recognized that the condition of the skeleton—which came to be known as Doreen—and that of the row of pots that had fallen from a shelf in the adjacent room, indicated neither murder nor burial: "All of us who actually dug up Doreen were in agreement: This looked very much like an earthquake."

Of course, there was another possibility. Doreen could have been killed by attackers, and later—perhaps years later—the wall fell on her bones and crushed them. Wanting to investigate further, Stewart consulted an expert in bone fracture analysis, Dr. Patricia Smith at Hadassah Medical Center in Jerusalem. She found evidence of a particular kind of breakage, a spiral fracture that only occurs in fresh bone. Her analysis indicated sudden and massive crushing of

Doreen's entire body while her bones were still clothed in flesh. In other words, Doreen was either alive or very recently killed when she was crushed by the wall. Incorporating this evidence into his report from the excavation, Stewart gave the opinion that the destruction in Doreen's layer of Dor was caused by an earthquake.

Strangely enough, despite Stewart's report, archaeologist Ephraim Stern, who directed the Dor excavation where Stewart uncovered Doreen, described the same find in one sentence: "On the floor of the Phoenician city that David conquered . . . amid the other evidence of destruction was the complete skeleton of a woman whose head had been crushed by a stone, apparently a casualty of battle" (Stern 1993). This sentence appeared in the same journal issue as Stewart's description; in fact, Stern cited Stewart's article in his bibliography. Why, when Stewart and his helpers found the evidence for an earthquake so compelling, did Stern completely ignore that interpretation? Perhaps, like many archaeologists, Stern simply avoids speculating about earthquakes because of cautionary examples like that of Claude Schaeffer (see chapter 1), or maybe he has not yet experienced a major earthquake in his native Jerusalem and is insensitive to the genuine earthquake hazard in all of Israel.

The destruction layer of which Doreen was a part is estimated to be some fifty years later than the general conflagration uncovered by Stern. Was the burned layer the mark of the invasion that brought Doreen's people to live at Dor? If so, why did the invaders set fire to the city when they meant to occupy it immediately thereafter? Alternatively, could an earlier earthquake have caused the older destruction layer as well?

Andrew Stewart raises this question in his 1993 article. The evidence, in his view, is equivocal but suggests, based mostly on an abrupt change in pottery styles, that the post-conflagration culture differed from the one destroyed in the fire. This may argue for conquest and occupation, or it may simply support the idea of a sudden influx of imported house goods. If most of the pottery was broken in an earthquake, an abrupt change in style might follow. The lack of a historical record means we may never know.

Although we have no independent, written record of the earthquake destruction of Dor, Doreen's shattered skeleton can be connected with an earthquake through medical evidence. Furthermore, numerous valuables were found under the rubble, the sort of plunder conquering armies remove rather than bury. More intriguing is the written evidence from the Bible for an earthquake at Michmash, only about 90 kilometers away, around 1020 BC, as noted in chapter 3:

> And the garrison of the Philistines went out to the passage of Michmash . . . And there was trembling in the host, in the field, and among all the people; the garrison and the spoilers, they also trembled, and the Earth quaked: so it was a very great trembling. . . . So the Lord saved Israel that day; and the battle passed over unto Bethaven. (1 Samuel 13:23–14:23)

The dating of Doreen's stratum of the ruins at Dor is based on an empirical clock developed from evolving pottery styles. Because this clock is only loosely correlated to biblical chronology, it is possible that the Michmash earthquake account is contemporaneous with either the destruction of Doreen's layer at Dor or with the earlier ash layer.

If the earthquake that killed Doreen was, in fact, the same earthquake described in the book of 1 Samuel, and if the biblical account is accurate, then this earthquake must have occurred during the day, as the troops were moving in the field of Michmash. The residents of Dor would have been going about their daily business. If the residents were awake and alert, many would have been outside, and others may have had time to react to the earthquake and rush outside, so that most of the destroyed rooms contain no skeletal remains. So far, Doreen's are the only remains to be found beneath the rubble.

KOURION, AD 365

Contrast the earthquake at Dor, where only one buried victim was found thus far, with one that occurred in Cyprus, in the fourth century AD. In the ancient port town of Kourion, on the southern coast of Cyprus, archaeologist David Soren and his colleagues uncovered a truly heartbreaking example of earthquake casualties. This town, originally founded by the Greeks during the Late Bronze Age, was struck by a series of devastating earthquakes beginning around AD 340. The final destruction of the city may have occurred just around dawn, when most of the city was still asleep. Soren and James (1988) used accounts from the ancient writer Ammianus Marcellinus, as well as archaeological evidence, to place the earthquake just after dawn, on July 21, AD 365.

It is important to note that some researchers do not agree with Soren's conclusions. Fokaefs and Papadopoulos (2004), for example, do not attribute the destruction at Kourion to the July 21, 365, earthquake and its accompanying tsunami, but to "other events occurring near Cyprus between 342 and 375." However, I believe that the evidence uncovered by Soren and his colleagues points unmistakably to an earthquake, regardless of whether the exact date can be agreed upon. Throughout Kourion, entire homes and stables collapsed, trapping people and animals inside beneath the rubble. So many skeletons were found, and the site was so undisturbed, that, in describing the excavation, Soren said that he and his colleagues "felt like a rescue team arriving 16 centuries too late" (1988, 39).

One room revealed a man in his thirties, crushed under rubble with his hands spread out to cover his head (Figure 5.3). Another room, apparently a stable, held the bones of a mule still tethered to a stone trough. Intertwined with the animal's skeleton is that of an adolescent girl, caught perhaps trying to calm her frightened beast. The most moving find, however, was the discovery of a family (Figure 5.4)—a man, a woman, and their eighteen-month-old child—trapped under their collapsed home (Soren and James 1988, 6). The three apparently had little time to realize what was happening, and their skeletons seem caught in a pose of sleep. The woman's arms are curled protectively around the bones of her toddler, with the man's arm flung over them both. Blocks from the ceiling, some weighing as much as three hundred pounds, had crushed all three.

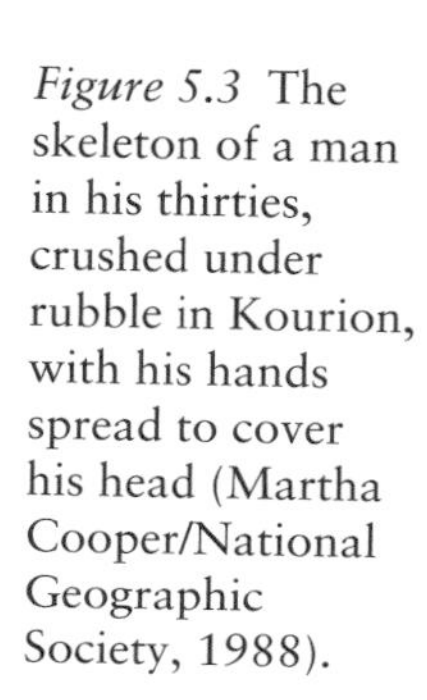

Figure 5.3 The skeleton of a man in his thirties, crushed under rubble in Kourion, with his hands spread to cover his head (Martha Cooper/National Geographic Society, 1988).

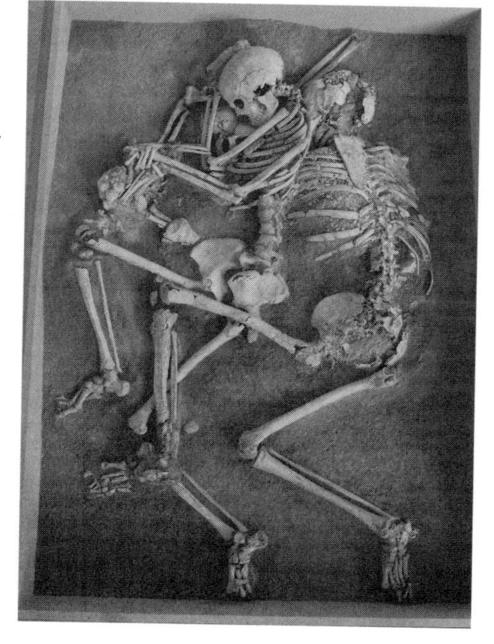

Figure 5.4 Skeletons of a family—a man, a woman, and their one-year-old child—trapped under the collapse of their home during the AD 670 earthquake in Kourion, Cyprus (Martha Cooper/ National Geographic Image Collection).

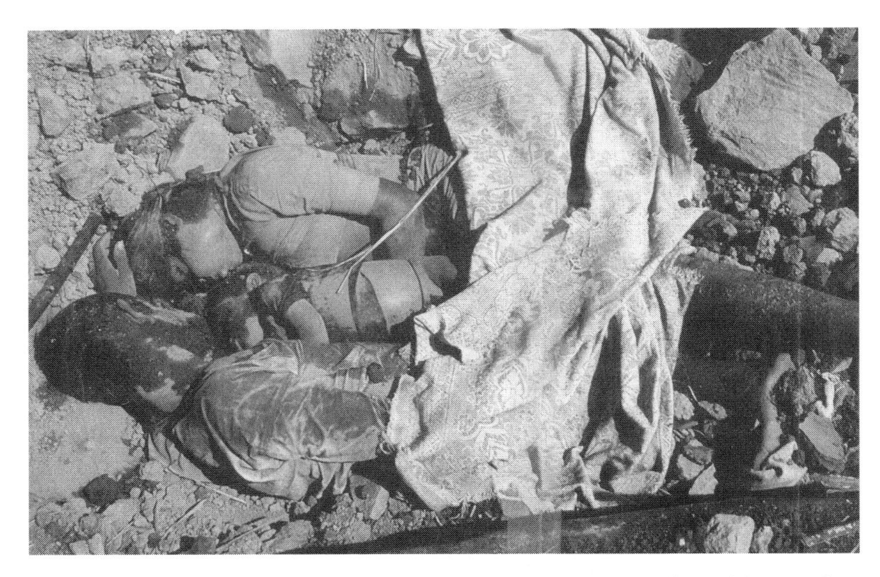

Figure 5.5 A mother, father, and child found dead in their collapsed home after an early-morning earthquake that rocked Killari, India in 1993 (Dieter Ludwig/ SIPA PRESS).

The horror of this find becomes even starker when compared to a similar scene from the 1993 Killari, India, earthquake, sixteen hundred years later (Figure 5.5). In an eerie echo of the past, a mother, father, and child were found dead in their collapsed home after the early morning earthquake. Just like the ancient family from Kourion, this young family had little chance to react.

PETRA, AD 363

The previous chapter described the devastation that the earthquake of 363 AD caused to the freestanding buildings of the city of Petra in Jordan, but there were also human casualties. Rolf A. Stucky of Basel University, the former director of the Petra excavations, unearthed the crushed skeletons of a woman and child in a Roman house in that city (Figure 5.6). With them was a cache of coins from a belt purse worn by the woman, one of several coin caches

Figure 5.6 Evidence of an ancient earthquake at Petra. Archaeologists of the University of Basel concluded that the skeletons of a woman and child found in Room 1 were a result of the May 19, 363 earthquake (from Stucky 1990).

found at the site over the years, and these coins enabled Stucky to place the likely destruction date at 363.

PELLA, AD 749

Nearly four hundred years later, an earthquake destroyed Pella, another town in Jordan. This was the same earthquake that, in AD 749, destroyed the Roman town of Bet Shean just across the Jordan River from Pella, as described in the previous chapter. A great many skeletons of animals and humans were found (Figure 5.7), crushed in the collapsed buildings of the city. That so many skeletons remain implies that, as in Kourion, the survivors were unable to recover the dead bodies for burial. Perhaps the collapse was so massive that they could not remove the rubble, or maybe there were so few survivors, and they were so dispirited, that they simply

Figure 5.7 An earthquake victim unearthed inside the colonnaded hall in Area IX, north of the Civic Complex Church in Pella, Jordan (from Smith 1982).

abandoned their homes and moved to what they hoped would be safer ground.

Also possibly affected by the same earthquake (although some scholars propose an earlier date) was the ancient site of Jerash (Gerasa), a Byzantine city situated some 34 kilometers north of Amman, the capital of Jordan (Amiran, Arieh, and Turcotte 1994; Browning 1982). Today Jerash is littered with the remains of theaters, colonnaded streets, temples, and other buildings and walls. Excavations of the Umayyad residential quarter here exposed a skeleton beneath the rubble (Figure 5.8). The destruction was so

Figure 5.8 Victim of the AD 749 earthquake, found in the devastated Umayyad housing quarter at Jerash (Courtesy Al Kubta Collection, 1988).

great that most of the city's inhabitants ultimately abandoned the city afterward.

GREECE, CA. 1190 BC

Several sites in Greece show signs of catastrophic destruction around the end of the Bronze Age. It is difficult to exactly correlate the dates of destruction for the separate sites, but many seem to have been destroyed during a period known as LH (Late Helladic) IIIB. The relative ages of the sites are correlated by the pottery styles that were prevalent in the destruction layers. As I understand it, this is not a terribly precise method, because some sites may be more current in their styles than others; moreover, other factors may influence pottery usage. Therefore, one cannot say with certainty whether the following five Greek sites were destroyed simultaneously or just within a relatively short span of years. (As I discuss in chapter 8, many regions of Greece are susceptible to quick successions of major quakes, called *earthquake storms*, so either scenario is possible.)

In any case, in each location, excavators have found skeletons crushed beneath fallen walls and collapsed buildings, and other evidence is indicative of earthquakes. This interpretation is quite controversial, partly because the destruction in Greece seems to be part of an even larger pattern of destruction throughout the Mediterranean region at about the same time. Ascribing this disaster to earthquakes was what cost Claude Schaeffer his reputation. Nevertheless, the skeletal evidence presented in this chapter corroborates earthquake damage in at least five sites. Later, I return to the issue of the wider-spread destruction.

Mycenae

In a building known as the Southwest House of the Citadel in Mycenae, destroyed sometime in LH IIIB, the north wall of one room was found collapsed, with the skeleton of a young man crushed under the burned debris. The south wall of the room, interestingly enough, had also collapsed, blocking the door to an adjoining room. Huge pieces of plaster from the walls of the building had also fallen, apparently before the walls themselves collapsed (Mylonas 1970, 1971; Iakovidis 1986).

In a house destroyed in the LH IIIB2 period, 200 meters north of the Citadel, the basement was filled with stone rubble, as well as the skeletons of three adults and a child crushed beneath them. The walls of several rooms were found fallen down the slope, and other walls were found leaning outward (Mylonas 1975; Iakovidis 1986; Mylonas-Shear 1987; French 1996; Maroukian, Gaki-Papanastassiou, and Papanastassiou 1996).

In another house, "the skeleton of a middle-aged woman whose skull was crushed by a falling stone was found in the doorway between the main room and the anteroom." The body was then buried by the debris of the house, which included smashed vessels and a chimney pot lying on the floor (Figure 5.9). Furthermore, the

Figure 5.9 Skeleton found in the doorway of Room 5, in House I of the Panagia group at Mycenae (from Mylonas-Shear 1987; with permission of the University of Pennsylvania Museum of Archaeology and Anthropology).

excavators noted "the collapsed state of the doorway leading into the house and the condition of the south wall of Room 2 where the preserved portion of that wall was found leaning outward, toward the south" (Mylonas 1962, 1963, 1966; Mylonas-Shear 1969, 1987). In this case, Mylonas concluded:

> House I was suddenly destroyed, but not by fire. The pile of stones found all over its area, the smashed vases with all their pieces in place under stones, the lack of burned remains, the discovery of a female skeleton in the doorway of its main room with skull broken by fallen stones, all seem to indicate that House I was destroyed by earthquake shortly before the middle of LH IIIB.

Another building, called the House of the Oil Merchant, was probably destroyed in late LH IIIB1 or early LH IIIB2; it showed signs of destruction that Wace and others believed to be the result of "enemy action" but that Iakovidis thinks "portray the results of a strong earthquake followed by fire rather than the disarray caused by looting" (Wace 1951; Wace et al. 1953; Iakovidis 1986).

Tiryns

The late Klaus Kilian, excavator of Tiryns, long argued that the site was destroyed at the end of LH IIIB2 (ca. 1190 BC) by an earthquake that also affected several other sites in the Argolid, such as Mycenae (cf. Kilian 1980, 1981). Other archaeologists such as Iakovidis (1986) agree with this assessment. Zangger (1991, 1993, 1994) recently suggested that a catastrophic flash flood that buried parts of the lower town of Tiryns (outside the Cyclopean walls) up to 5 meters deep, may be related to this earthquake, perhaps owing to the destruction of an ancient dam. Kilian (1996), in a paper published posthumously, reported that "the evidence consists of building remains with tilted and curved walls and foundations, as well as skeletons of people killed and buried by the collapsed walls of houses. . . . [A] comparative study of buildings that have been affected by earthquakes in the last 100–200 years supports our conclusions that the observed deformation of excavated buildings are of seismic origin." Kilian also suggested that another earthquake had damaged Tiryns during the LH IIIC period.

Within a large complex inside the Acropolis dating to the time of the last palaces, the skeletons of a woman and a child (Figure 5.10) were found "buried by the walls of Building X (Kilian 1996)." Kilian adds that, within Building VI,

> A high wall was transformed into a mass of rock. . . . [T]he walls on the terrace and on the other side of the corridor are tilted downhill (westwards) and uphill (eastwards) respectively, that is, in a direction opposite to that of a possible slope-movement. . . . Such antithetical tilting of nearby walls is not the result of landslides but of seismic disturbances.

Figure 5.10 Skeletons of a woman and child buried by the fallen walls of Building X at Tiryns (from Kilian 1996).

In an early LH IIIB (ca. 1300–1260 BC) house, "a skeleton was found beneath the fallen walls" (Kilian 1996, 65). Based on these skeletons, and on deformed walls from several other layers, Kilian believed that earthquakes were a recurrent theme at Tiryns, and his was not the only investigation at Tiryns to find skeletal remains beneath collapsed walls. In 1956, Verdelis excavated fill near the fortification wall with "LH IIIB sherds, fragments of wall paintings, and remnants of burnt wooden beams." He discovered "two skeletons, evidently not burned but killed and covered by the fallen debris" (Verdelis 1956; Mylonas 1966).

Midea

Another site in this same region is Midea, situated between Mycenae and Tiryns. The excavators of Midea suggest that destruction at the site in LH IIIB2 (ca. 1190 BC) was caused by an earthquake. They cite, in particular, "collapsed, distorted, curved walls" as well as a skeleton found under collapsed debris (cf. Åström and Demakopoulou 1996, 37, 39; Shelmerdine 1997, 543; and Demakopoulou 1998, 227).

In one of the rooms in the area of East Gate, in an LH IIIB2 context, "the skeleton of a young girl was found, whose skull and backbone were smashed under fallen stones" (cf. Åström and Demakopoulou 1996, 39; and Demakopoulou 1998, 227). In addition, buildings located inside the Acropolis to the left of the West Gate were destroyed by fire and contained large stones and nearly complete mudbricks that had fallen from the walls of the rooms and from the collapsing fortification wall (Åström and Demakopoulou 1996, 38–39).

Thebes

Excavations of the Kadmeia citadel in Thebes in 1980 yielded evidence of a destruction in late LH IIIB1, which excavators attributed to "a sudden earthquake succeeded by fire of long duration" (Sampson 1996). Previous excavations also hinted at a destruction caused by an earlier earthquake in the LH IIIA2 period (Symeonoglou 1987).

Part of a palatial workshop in the eastern wing of the Kadmeia yielded "a thick destruction layer . . . more than 1 meter thick . . . Unbaked mudbricks coming from the fallen walls of the building were found in different levels; some of them were later baked by the fire. The destruction was immediate" (Sampson 1996, 114). Within this general destruction layer, a human skeleton was found 0.70 meters above the floor in Room I of the building:

> The skeletal remains were within the destruction layer; since it was overlain by a much harder stratum of erosion . . . it cannot postdate the destruction layer. The remains were found well above the ground floor which suggests that the person was on the first floor at the dreadful moment of the destruction, could not escape and was finally trapped among the ruins. . . . A careful anthropological study suggests that the

> skeleton belongs to a young female of about 20 to 25 years old and 1.55 meters in height. Injuries are evident on the skull, but what caused death was a fatal depressive detaching fracture in the middle of the cranium vault . . . very likely produced by a very violent blow from a sharp structural material—probably a roof beam—which hit the woman suddenly. (Sampson 1996, 114)

The Menelaion

At the Menelaion in Greece, which was located near the site of later Sparta, the British archaeologist Hector Catling noted a monumental terrace wall that had collapsed during the LH IIIB2 period: "When the terrace wall collapsed it did so suddenly and unexpectedly. This is implicit in the 1978 discovery of a human skeleton trapped in what is now seen to have been the terrace collapse" (Catling 1981).

The deaths of all these earthquake victims must have been unspeakably tragic when they occurred. We can only imagine how long their surviving relatives searched for their bodies or how they felt when they finally gave up, consigning their skeletons to the rubble. For archaeology, however, these events are remarkable windows on the past, offering us a glimpse of ancient people, not prepared for burial and the afterlife but caught unawares in a last moment of ordinary living. The woman with her purse of coins from Petra, the sleeping family from Kourion, the convalescing Neanderthals from Shanidar—these were people as their contemporaries knew them. They grant us a more human insight into ancient life than we can obtain from the self-conscious pomp and splendor of elaborate tombs or carefully chosen grave goods.

Perhaps the only event that better serves as a time capsule of ancient life is a volcanic eruption like the one that entombed Pompeii. Such eruptions, however, are much rarer than earthquakes. The earthquake rubble of millennia has preserved for our eyes many secrets of ordinary ancient life that might otherwise remain unknown.

CHAPTER 6

Qumran and the Dead Sea Scrolls

DESTRUCTION THAT PRESERVES?

Instead of learning about earthquakes from archaeology, we ended up learning about archaeology from earthquakes.
—Amos Nur, Lecture to the Seismological Society of America, 1997

The ruins of Qumran, situated amid the limestone cliffs on the northwest shore of the Dead Sea (Figure 6.1), are the focus of considerable mystery and controversy. Scholars have long debated nearly everything about this site, including the identity of the people who lived there, why they abandoned it, the dates of occupation, and the significance of the architecture and artifacts. In my opinion, the Judean earthquake of 31 BC not only serves as a useful marker in the archaeological layers in Qumran but also played a major role in its confused history and affected the surrounding caves. There is overwhelming evidence that the quake heavily damaged Qumran (Allegro 1964; Milik 1959; de Vaux 1973), disrupting its water supply and forcing residents to leave for some time. The shaking probably also collapsed several of the caves near the settlement, perhaps leading to the abandonment of what eventually became the archaeological find of the twentieth century. By destroying the buildings and caves of Qumran, the earthquake of 31 BC may have played a part in preserving an

Figure 6.1 Aerial view of Qumran and some of the caves in the cliffs below.

incalculable treasure, a contemporary written record of one of the world's most pivotal periods of religious and cultural history.

Qumran is a place that has spawned political tricks and conflicts between Israel and Jordan, and it has sparked heated debate among archaeologists, historians, and Christian and Jewish apologists. The ruins of Qumran went unnoticed and unexcavated, however, until a Bedouin shepherd boy chasing his goat made an amazing discovery in 1947.

According to the prevailing story, one of the boy's goats escaped from the flock and climbed up into the cliffs of Wadi Qumran. The boy followed and, when he stopped to rest, saw a small opening high in the cliff. He threw a rock into the opening and was startled

Figure 6.2 A fragment of the Book of Psalms containing parts of chapter 15 and the beginning of chapter16 found in the Cave of Letters, possibly inscribed by the Essenes (From *Bar-Kokhba* by Yigael Yadin, copyright © 1971 by Yigael Yadin. Used by Permission of Random House, Inc.).

to hear not the thud of rock on packed earth but the unmistakable clink of fired pottery. Eventually the boy entered the cave and found it full of jars, at least one containing a cache of parchment scrolls, browned with age, and wrapped in linen cloth. This was the first of several caves in the region where ancient scrolls or scroll fragments were discovered (Figure 6.2), sometimes in jars, sometimes simply wrapped in linen. The scrolls, known collectively as the "Dead Sea Scrolls," were first dated to the late first century BC by comparing the writing style with other sources. The date was later revised to the first century AD based on a confused record of coins and pottery from the excavation of Qumran, but this revision, too, has been cast into doubt. Parchment from some of the scrolls has been dated using Carbon-14 measurements, but the resolution of this method is insufficiently precise to settle the matter. The range of probable dates, though centered precisely on the earlier date, also includes the later one.

Only in 1951 was Qumran itself excavated, and by then most archaeologists already considered it the most likely source of the Dead Sea Scrolls. Today the ruins seem incredibly isolated, surrounded by desert in an area where the temperature soars well over 100°F in the afternoon, and where it is difficult to imagine maintaining a functional community. In its heyday, however, Qumran was neither as isolated nor as parched as it is today. It was only nine miles from Jericho and two miles from the spring of Ein Feksha on the shore of the Dead Sea, where a small settlement thrived and grew dates during the time Qumran was inhabited. Lacking a freshwater spring of its own, Qumran had an elaborate system of aqueducts to funnel seasonal rainwater into its huge cisterns for storage. We cannot know with certainty the number of people who lived there, but Qumran's cemetery, with nearly a thousand graves, attests that this was a thriving community at some point.

Qumran is thought by many to have been a headquarters of a sect called the Essenes, although this, like everything else, is disputed. The Essenes, probably an ultraconservative Jewish sect, are believed to have left Jerusalem during the Roman occupation because their strict interpretation and implementation of Jewish law made living in the Holy City nearly impossible for them. Because Jerusalem was the city of the Temple, the Essenes believed that the entire city was subject to extreme purity laws. By their interpretation of the law, they were forbidden to perform many necessities of daily life within the city walls. They even had to leave the city to defecate.

The general populace of Jerusalem, not surprisingly, did not follow these rules. Not only did they interpret the purity laws differently, but they had also begun to adapt Jewish religious observances to the now dominant Roman calendar. The Essenes, offended by what they saw as a rejection of Jewish values, believed they could become tainted by association. They moved into isolation in small cities and towns in the wilderness, where they could live according to the law as they saw it, without being touched by the iniquity of the times. They perceived the Roman cultural invasion as a sign of impending doom, and many chose to live apart, copying and interpreting the holy books while awaiting an eventual Apocalypse, which they were certain was imminent.

WERE THEY CHRISTIANS?

Another enduring debate among scholars of the Dead Sea Scrolls concerns the connection between the Essenes and the early Christians. Some Christian theologians have argued that the Essenes were, in fact, early Christians. The Essenes, after all, emerged approximately at the beginning of the Christian era, and their message was fervently Messianic. Some historians believe that John the Baptist may have been an Essene himself or else closely allied with the sect. However, the Essenes continued to wait for their messiahs—they expected two of them, not one—long after the ministry of Jesus, and their emphasis on strict adherence to ritual and law, which even early Christianity discounted, argues against the two movements being one group.

In 1948, Professor Eliezer Lippa Sukenik from the Hebrew University of Jerusalem was the first scholar to suggest Qumran's link with the Essenes mentioned in the ancient sources. The French Dominican priest Father Roland de Vaux and his team accepted the Essene origin of Qumran as a working assumption, which then became fixed in public perception. The only historical justification for this assumption comes from Pliny the Elder (1938), who, in *Natural History* (ca. AD 77), describes the Essenes as living in the desert near the Dead Sea, but other contemporary sources, including Philo of Alexandria (30 BC–AD 45, cited in Magness 2002, 40) and Flavius Josephus (1991a, 1991b), distribute the sect more widely. The uncertainty about the Essenes themselves is no greater than the uncertainty that has plagued the Dead Sea scrolls up to the present day.

Many of the scrolls could certainly be the work of the Essenes, or a similar, perhaps less pacifist, conservative sect. Some of the texts are devoted to describing the strict rites of ritual purity that were required for attending meetings with the sect and for dining with them and becoming a full member. This still does not prove, however, that Qumran was the home of the Essenes. Some scholars wonder whether, rather than being a library of Qumran, the caves around the site might have been repositories of scrolls brought from many parts of Judea, or from Jerusalem itself when the Temple was destroyed by the Romans in AD 70. In that case, linking the inhabitants of Qumran too closely with the writers of the scrolls would be a mistake.

A number of scholars argue that Qumran could not have been a sectarian monastery. It has been variously described as a trade center (Crown and Cansdale 1994), a winter villa (Donceel and Donceel-Voute 1994), and a fortress (Golb 1995). The large inventory of pottery found at Qumran, along with a kiln, does suggest some commercial activity. The residents produced pottery, probably some of the items for use elsewhere, including possibly in Jericho. However, no major commercial route is known to have passed through the ancient site, and its use as a hub of commercial activity seems unlikely.

The University of Chicago's Norman Golb (1995) maintained that the site functioned as a fortress during stabilization of the Hasmonaean state:

> There was nothing whatever at . . . Qumran to attest to its being a monastery, a place where monks or other notable sectarians lived, or a center where scholarship, intense writing activity, or the copying of books was pursued. The evidence showed that identification of the site as the home of Pliny's Essenes was untenable.

G. Lankester Harding, in his book *The Antiquities of Jordan* (1959), claimed that the site was a fortress starting around the seventh century BC but was abandoned until the second century BC when new inhabitants settled there. However, Curator Emeritus Magen Broshi of Jerusalem's Shrine of the Book, where several of the Dead Sea Scrolls are housed, and the University of Haifa's Joseph Patrich both reject the idea that the site was ever a military fortress. "This seems an unlikely explanation," Broshi stated, "as the site is of inferior strategic value, and the flimsy walls of the complex could not have had military value." The notion that the place was a fortress seems based entirely on the presence of the thick-walled tower at one corner of the complex, but this could have served a defensive purpose for an isolated community or perhaps had a religious significance in accordance with the Essene belief in the coming Apocalyptic battle.

Many of the excavators, particularly Roland de Vaux, had possibly inappropriate preconceptions about what monastic Essenes would have been like, extracted from their own knowledge of Roman Catholic monastic experience. For instance, none of the scrolls associated with the laws of the community explicitly prohibits the presence of women, but nearly all the excavators assumed that any Essene monastic community must have been made up only of men.

However, recent surveys of the cemeteries associated with Qumran seem to lend credence to the monastery theory. The cemetery is rigidly organized, the tombs that date from the Second Temple Period are nearly all of similar layout, and there is a typically monastic lack of jewelry and other grave goods. If it were, as some have argued, a military cemetery, there should have been a preponderance of battle injuries or other signs of violence in the bones there. Instead, the only damage is consistent with that seen in other cemeteries of this age, damage that could have accrued over time after the bones were buried (Zias 2000). For an exhaustive discussion of the fascinating archaeology of Qumran and the competing theories about its inhabitants, see the book by Jodi Magness (2002), *The Archaeology of Qumran and the Dead Sea Scrolls*.

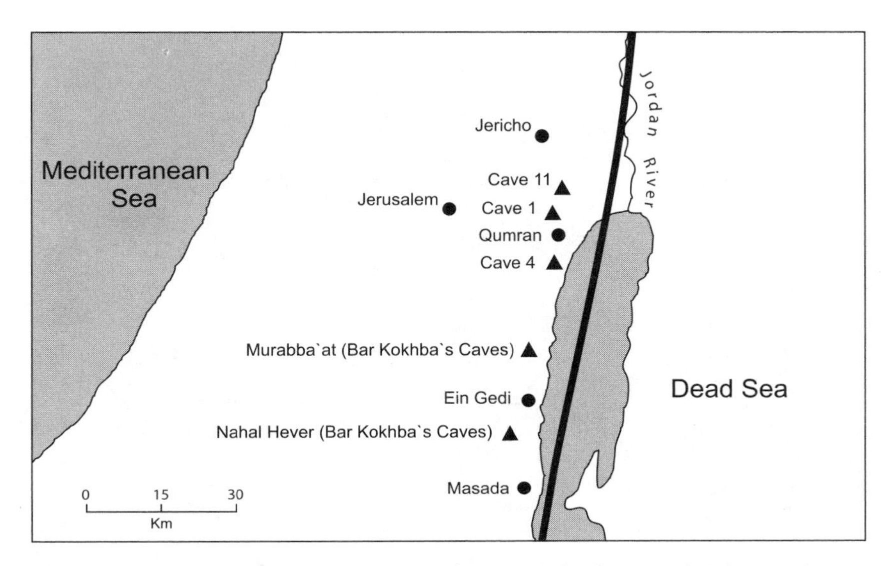

Figure 6.3 Map showing the locations of caves and other sites by the Dead Sea.

Putting aside for the moment whether the inhabitants of Qumran were Essenes, whether the Essenes were early Christians, and whether the scrolls were the labor of Qumran, one event is not disputed: in or around 31 BC, an earthquake severely shook Qumran and much of Judea. Qumran sits on a very active fault zone, which is currently in a seismic quiet period (Figure 6.3). The city was badly damaged by movement during the earthquake (Figure 6.4), and many of the caves where the Dead Sea Scrolls were found show evidence of earthquake damage. Josephus (1991a) describes the rupture of the central water cistern in Qumran, which forced the inhabitants to abandon the town for at least one winter and perhaps for as long as several decades. The two-thousand-year-old fault rupture that offsets the stairs of the cistern, unearthed by archaeologists some years ago, appears as fresh as if the earthquake had happened yesterday (see Figure 4.4).

If Qumran was indeed the home of the Dead Sea Scroll sect, which I believe is most likely, then their way of life, as chronicled in the scrolls, depended on living apart from the unclean and keeping themselves ritually purified, a task requiring great quantities of

Figure 6.4 Signs of destruction: Plates, bowls, and goblets discovered in one room at Qumran, with dozens of vessels piled one on top of one another. This room probably served as a crockery near the assembly room, which may have been a dining room (Courtesy Israel Antiquities Authority).

water. Furthermore, much of their faith rested on members' assurance of their own preordained righteousness. The destruction of their city, and especially the disruption of the water supply needed for their religious practices, must have been a blow to their view of the world. Nonetheless, most archaeologists agree that the same group, or a group similar enough to put the buildings to the same uses, reoccupied Qumran at some time after the earthquake, clearing away debris from some parts of the complex and bringing part of the water system back into use. The cistern in Figure 4.3, however, was never repaired. The settlement then came to a violent end some time in the middle of the first century, and Roman-style arrowheads litter the final layer of debris.

Because of the Arab-Israeli conflict, excavation of Qumran has been sporadic and the quality of the archaeological work there has suffered during some periods. However, the situation there seems pristine and controlled compared to the "excavation" of the scroll-bearing caves throughout the area. The Dead Sea Scrolls, like written finds in many other sites, forced reinterpretations of much of what had been considered common knowledge in the field of archaeology. Perhaps more important, the scrolls offer a glimpse of the Second Temple Period, an almost unknown period in Jewish history that preceded the rise of Rabbinical Judaism, which eventually completely dominated Jewish thought worldwide. Given the importance of the scrolls, it is a lasting shame that the tremendous turmoil, trickery, and mistakes surrounding their discovery and interpretation have destroyed much of the supplementary information that the caves originally could have revealed.

TRICKS

In all the uncertainty shrouding the discovery of the scrolls, one fact stands out: the first scrolls, and indeed almost all of them, were found by Bedouin shepherds. According to the most popular story, described by Allegro (1964), Magness (2002), and others, a young Bedouin named Muhammad Adh-Dhib stumbled upon the first of the scroll-containing caves while climbing after his stray goat. The story goes on to tell how the boy retrieved a few leather scrolls for his elders. The elders, ignorant of their worth, sold the scrolls to Khalil Iskander Shahin, known as Kando, the owner of a Bethlehem cobbler shop. Allegro wrote that Kando was also unaware of their value when he bought them, and he hoped to find some use for the leather in his cobbler shop. Later, upon closer inspection of the writing on the leather, Kando began to suspect that the scrolls were extremely ancient and sought a buyer for them on the antiquities market.

This is a quaint tale but is certainly mostly fiction, as flimsy as the fragile, crumbling leather Kando supposedly bought to use for shoes. Despite the portrayal of the Bedouins as naïve goat herders with no idea of the significance of their find, tribesmen in the region had been supplementing their incomes since the 1920s by selling ancient artifacts they found in the desert. The players in this story, like many of their kin, may have been archaeological treasure hunters, who well knew that dealers in antiquities would be interested in their find. In any case, the Bedouins had apparently been poking about in caves for a long time and were familiar with the cave geology of the Judean wilderness. Kando obviously suspected that the scrolls were valuable before he offered any money for them, and he soon organized expeditions to find more. To this day, despite repeated expeditions led by professional archaeologists from Jordan and Israel, most scroll finds in the region have been made by the Bedouins.

The Bedouins approached Kando with the first of the scrolls just before the outbreak of the Arab-Israeli war in 1948, and Kando would long serve as their representative to various people and agencies interested in the scrolls. He sold the four fragments to Athanasius Yeshua Samuel, a bishop in the Syrian Orthodox Church where he was a member, and then sent an expedition back to the cave despite military roadblocks in the area. On the eve of the war, Kando sold additional fragments to the archaeologist Eleazar Lippa Sukenik, at the Hebrew University of Jerusalem. Meanwhile, Bishop Samuel, knowing that the continued hostilities would make it difficult for him to get top dollar locally for the four scrolls he had purchased, placed an advertisement in the *Wall Street Journal* on June 1, 1954, seeking buyers for his scrolls (Figure 6.5). Over the years, antiquities speculators have smuggled uncounted fragments of the Dead Sea Scrolls to black markets around the world, and many—perhaps most—of the discoveries have not reappeared to this day.

CONFLICTS

As Allegro (1964) describes, the Bedouins who found the scrolls did not want to reveal the location of their cache either to their dealers and buyers or to their own Jordanian government. Gradually it became known, however, that scrolls were being discovered in caves near the Dead Sea.

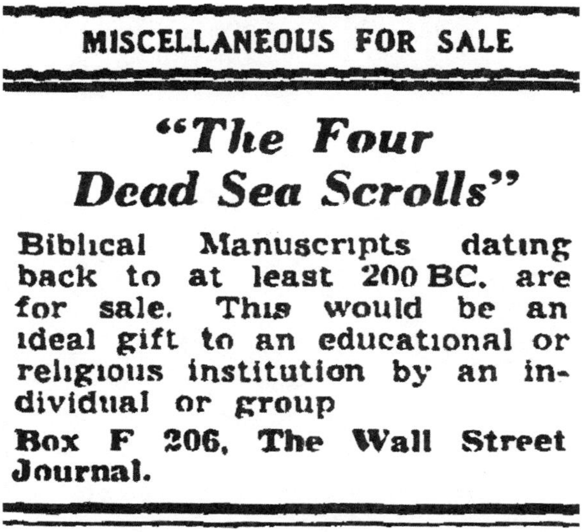

MISCELLANEOUS FOR SALE

"The Four Dead Sea Scrolls"

Biblical Manuscripts dating back to at least 200 BC. are for sale. This would be an ideal gift to an educational or religious institution by an individual or group

Box F 206, The Wall Street Journal.

Figure 6.5 Samuel's advertisement in the *Wall Street Journal* for public sale of the scrolls.

By the end of the 1948 Arab-Israeli war, the area where authorities suspected the scrolls had been found had become Jordanian territory. Soon, officials of the Jordanian government began their own search, looking both for the caves the Bedouins had found and for other caves that might contain similar finds. They were dismayed that what might turn out to be the most important archaeological relics ever discovered in their country were being sold to the highest bidder. Smuggling a national treasure was a crime, but the Bedouins could never be caught with any quantity of the merchandise; they carefully divulged only a scroll or two at a time. The government knew that cracking down on its few Bedouin contacts would only destroy any chance to buy the bulk of the scrolls for the country's own museums. Despite numerous searches and discoveries of "new" caves, officials repeatedly found that the Bedouins had preceded them, removing most of the scrolls and irrevocably scrambling the archaeological context of the sites.

The price demanded by sellers soared as the shepherds and their intermediaries realized there was an eager international market for the scrolls. Israel was particularly interested, since the ancient writings chronicled a missing link in Jewish religious history. Eventually, however, rumors surfaced that some of the finds came from caves actually inside Israeli territory.

It now seemed that Israel was losing its national treasures. Therefore, in 1953, the Israeli government arranged for its own expeditions to some of the Judean desert caves on the Dead Sea, north of Masada and close to the Jordanian border. The group dispatched to the cave which eventually became known as the Cave of Letters consisted of the archaeologist Yoharan Aharoni, a few other professionals, some members of the Israeli army, and many volunteers. One of these volunteers was Israeli kibbutznik Baruch Safrai.

NEW DISCOVERIES

Safrai (1993) recalls his involvement in the arduous expedition in an article published in *Biblical Archaeology Review*. The Cave of Letters is located in the face of a canyon wall, almost 460 meters above the base of the cliff and 15 meters from the top. To enter the cave each morning, the excavators had to climb down a rope ladder from the top of the cliff to a ledge below the cave, and then up again to the cave's narrow entrance. Safrai, a fisherman at one time and familiar with knots, was in charge of tying the ladders and other rope work. Once in the cave, the excavators had to deal with pitch darkness, a shortage of fresh air, and a ubiquitous thick coating of bat guano. The floor of the cave was a mass of boulders and rubble that had fallen from the ceiling at some unknown time in the past—the distant past, judging from the guano. The extreme dryness of the cave meant that any digging or shifting of the rubble quickly stirred up choking clouds of guano dust that would smother lanterns and make breathing difficult.

By the faint light of their lamps, the excavators found only disappointment. They had hoped to uncover scrolls like those found around Qumran but, besides a few broken shards of pottery and some wood, stone, and cloth artifacts, they found only a few empty Jordanian cigarette packs, evidence that the Bedouins had been there ahead of them. The team searched the surface of the rubble but found little; there were no scrolls. Safrai offered to try to shift some of the boulders on the floor, hoping that the fall had occurred after the period of the scrolls, and that something of value might lie beneath the rubble. Tunnels dug earlier, presumably by the Bedouins, indicated that others, too, had thought this approach valuable. After moving a few of the smaller stones, the enthusiastic young Israeli wormed his way headfirst down among the larger rocks, holding a flashlight in front of him. Approximately a body-length into the rubble, he made a startling discovery.

According to Safrai (1993), pinned on its side beneath one of the rocks was a skeleton. Safrai described how the "position of the skeleton under the heavy boulders clearly indicates that this was no ordinary burial; the skeleton lay sprawled out, apparently crushed when the roof of the cave collapsed." The skeleton was still clothed in the remains of a white garment, tied at the waist with a rope belt. Safrai extricated himself from the rocks and ran to report his discovery to Yohanan Aharoni, the archaeologist in charge of the expedition. Returning with Aharoni and some padded boxes, Safrai again wriggled in among the boulders to describe the scene to the expedition leader; he then tried to retrieve the skeleton. Unfortunately, he was able to grasp only part of the rope belt and robe, which went into a box to be shipped back to Jerusalem, but the bones lay beyond his reach. The robe and belt fragments were never seen again. No one from the expedition knows what happened to the box, but apparently it never reached Jerusalem. Safrai's description of the clothing, however, was tantalizing.

The robe Safrai described does not rule out the possibility that the skeleton was that of an Essene. According to Josephus (1991a [AD 75]) white clothing was significant to the Essene community: "They think to be sweaty is a good thing, as they do also to be clothed in white garments." Josephus provided additional information about the Essene garb when recording the steps through which an initiate must go to join the community: "If any one hath a mind to come over [to] their sect, he is not immediately admitted, but he is prescribed the same method of living which they use, for a year, while he continues excluded; and they give him a small hatchet, and the forementioned girdle, and the white garment."

Tufts University archaeologist Jodi Magness (2002) also found undyed linen fragments in the caves of Qumran and noted that, among the Essenes, only men wore white garments. Historical sources describe the Essenes as living in various parts of the Judean desert, not only around Qumran. Could Safrai have seen the skeleton of an unfortunate Essene, crushed beneath the weight of the cave's collapsed ceiling? If so, could there possibly be scrolls beneath the rubble in this cave waiting to be discovered, like the ones found near Qumran?

The earliest mention of Safrai's 1953 discovery was apparently by Rothenberg and Aharoni (1960) in their book, *In the Footsteps of Kings and Rebels*. Perhaps its late publication is why, when the archaeologist Yigael Yadin (1971) led another expedition to the cave in 1960, no one took seriously the assertion by the earlier expedition that a layer with evidence of habitation was beneath the rubble. Yadin made an amazing but unrelated discovery in the cave: letters from Shimon Bar-Kosiba, the legendary, mid-second-century leader of the second Jewish revolt against the Roman occupying force in the Holy Land, whose real name was not even known at the time; he was known by a messianic corruption of his name, "Bar-Kokhba," meaning "son of the star" (Figure 6.6). Yadin (1971, 60) summarized his dismissal of the possibility that any remains could lie below the rocks, below the level of the Bar-Kokhba rebels:

> The report of the 1953 expedition to this cave was our point of departure. As already mentioned, the most important conclusion reached by that expedition was that the large blocks or rock falls, covering the cave floor, fell from the ceiling after the last inhabitants had left the cave. As these blocks are quite immovable and as it was obvious that any use of explosives might cause further rock falls, we decided to ask the Israel Defense Forces for several pneumatic drills with which to break up the blocks in a less violent manner. Transporting the drills and bringing them up the rope ladder was a most difficult task, as was the actual

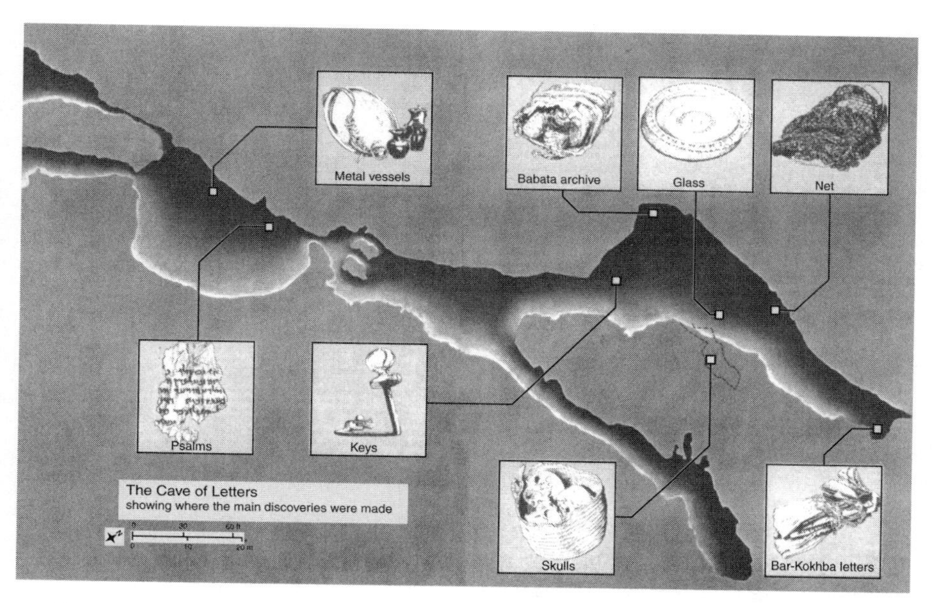

Figure 6.6 Yadin's findings in the Cave of Letters (From *Bar-Kokhba* by Yigael Yadin, copyright © 1971 by Yigael Yadin. Used by Permission of Random House, Inc.).

> breaking-up of the blocks later; these efforts ultimately proved fruitless, for the lack of oxygen in the cave stifled the combustion of the drills shortly after their activation and the exhaust fumes made breathing—hard enough as it was—virtually impossible. Luckily we determined after only a very short while that the large blocks of rock must have fallen *before* [emphasis in original] the period of Bar-Kokhba, since many areas of the ceiling were still covered with soot, indicating that there had been no substantial changes in the structure of the cave since ancient times. Therefore there was little sense in trying either to move or to break up these huge blocks of rock, *for no level of habitation could possibly lie beneath them* [emphasis ours].

The conclusion that the Bar-Kokhba rebels, who fled to the desert and hid in the remote hillside caves, dwelt atop the rubble is probably correct. This would account for the sooty ceiling. Yadin's subsequent conclusion, however, that no lower occupied level could exist, is completely unfounded; in fact, it is difficult to understand how he could have drawn that conclusion save by mere wishful thinking in light of the hardships he would face in trying to shift the rubble. Still, his expedition did recover a rich collection of remains from the Bar-Kokhba period from niches in the rubble, including the letters from Shimon Bar-Kosiba himself. Yadin's subsequent excavations, understandably, focused entirely on this period. The disappointing result of his assumptions, however, was that the excavation proceeded without regard for preserving possible lower strata in the cave. As we eventually learned, this made later work more difficult.

The collapsed ceiling in the Cave of Letters is not unique; many of the scroll caves show signs of rockfalls from their ceilings. The Cave of Letters, like others in the area, and like the distant but geologically similar Shanidar Cave, is very broad, with a relatively flat ceiling (Figure 6.7). The broad expanse of unsupported ceiling, combined with naturally occurring cracks and joints in the rock, makes the configuration unstable. Even a relatively mild earthquake can trigger collapse under these circumstances—but when did the collapse occur?

In 1991, I worked with my colleagues, Chris McAskill and Hagai Ron, to create a documentary video, *The Walls Came Tumbling Down: Earthquakes in the Holy Land*. We blended aerial shots of the region, archaeological excavations, geophysics, computer simulations, ancient art, and commentary from several biblical and historical sources to reveal the drama of earthquakes in this region, including the earthquake of 31 BC. Israeli educational television carried the documentary, and Baruch Safrai, now an elderly man, saw it. (Recall that Safrai had been the young archaeological volunteer who discovered the skeleton in the Cave of Letters.) Perhaps the ceiling had collapsed in the 31 BC earthquake, Safrai speculated, burying everything, including any scrolls and any people who would have been attending to them at the time. This collapse, then, would have happened more than two hundred years before Bar-Kokhba, and hence the confusion of Yadin's expedition.

Figure 6.7 The very narrow entrance of the Cave of Letters. Note the flat ceiling.

Safrai contacted us with his remarkable idea, and we eventually organized an expedition with a crew from Hebrew University to see what we could find. We went to the Cave of Letters in March 1996, armed with a map Safrai had drawn, and entered the cave as other teams had done before us, with ropes and rope ladders (Figure 6.8). We found the site just as Safrai and Yadin had described it (Figure 6.9), with enormous heaps of limestone blocks that had fallen from the cave's roof, completely burying anything underneath. We searched and dug, and some of the more agile among us tried to wriggle down into gaps in the rubble, but we were unable to find the skeleton the young volunteer had seen, and were also unable to reach the layer beneath the rockfall anywhere else in the cave. We feared that Yadin's excavation of the cave may have shifted rubble enough to plug the crevice Safrai had discovered, or perhaps the ever-present Bedouins had found it before us and removed anything of interest. In any case, the idea that a skeleton lay crushed beneath the rocks remained provocative but unconfirmed.

Figure 6.8 The author climbing down a rope ladder from the top of the cliff to enter the Cave of Letters.

Figure 6.9 A photo of Yadin's excavation in 1960, showing the enormous heaps of limestone rubble that had fallen from the cave's roof, completely burying anything underneath (From *Bar-Kokhba* by Yigael Yadin, copyright © 1971 by Yigael Yadin. Used by Permission of Random House, Inc.).

We hoped to see an expedition one day with the means to reach beneath the rubble.

In 1999 and the following years, the archaeologists Richard Freund and Rami Arav led just such an expedition. Baruch Safrai had also approached them regarding the skeleton he had seen in 1953. Unlike our expedition, however, theirs hauled into the cave two high-tech tools that would prove invaluable in their explorations: ground-penetrating radar equipment (to image the layers of debris in the cave) and a medical endoscopic camera, for peering into the inaccessible crevices beneath the boulders.

During their exploration of the cave, Freund and Arav discovered many artifacts (including coins) dated to the first century AD, some of them discovered in niches between the boulders. Although Freund (2004) concedes that the Bar Kokhba rebels may have brought these artifacts with them, they certainly do not rule out an earlier occupation.

The most exciting discovery, of course, was the refutation of Yadin's dismissal of a habitation layer beneath the rubble. Using their endoscope, the team rediscovered Safrai's skeleton, still with fragments of its tunic and belt. They also found bones of a second individual, trapped beneath the boulders. Clearly, someone had been in the cave when the ceiling collapsed. The endoscope also revealed many artifacts beneath the rubble, including fabric, combs, coins, and scraps of papyrus, though unfortunately without inscriptions (Freund 2004).

On our earlier expedition in 1996, in addition to exploring the Cave of Letters, we briefly surveyed other caves in the area, many of them also with collapsed ceilings. We know that the documented 31 BC earthquake displaced in-ground cistern walls in Qumran, besides causing widespread damage in most of Judea. As noted in other chapters, it is rare to see such ground displacement in earthquake ruins, since only a large earthquake will rupture the ground surface at all, and ground rupture is generally seen only right on the slipping section of the fault. The 31 BC earthquake has also been correlated to a layer of mixed sediment on the Dead Sea floor, one of the seismites described in chapter 2. All this evidence indicates that the shaking from the 31 BC earthquake would have been very severe in the Qumran region, probably strong enough to cause severe damage to caves with broad, unsupported, weak ceilings.

Freund (2004), however, obtained radiocarbon dates for some of the bones he found beneath rubble in the Cave of Letters, and these remains dated, at the earliest, to the first century AD. Thus, he proposed that an earthquake reported to have occurred in AD 115 was the most likely candidate for collapsing the cave. The Cave of Letters is far enough from Qumran that the 31 BC earthquake could conceivably have affected the two sites differently; I find it curious, however, that a thorough study of seismites in the Dead Sea (described fully in the next chapter) uncovered evidence of the 31 BC earthquake but no evidence of an earthquake around 115 AD. It seems unlikely that an earthquake large enough to bring down a ceiling that had survived the 31 BC quake would have left no record in the Dead Sea sediments. This question, however, needs more research and collaboration between archaeologists and earth scientists to sort out the dates, the layers of evidence, and the seismicity that might have influenced the cave.

Nevertheless, the seismicity of the area makes it highly likely that many of the caves in the Judean desert may have collapsed during or shortly after the period when the Essenes were active, perhaps while they were storing scrolls in the caves around Qumran. Note that archeologists disagree as to when the scrolls were stored; some believe that Jews fleeing the Roman attack on Jerusalem around 70 AD, and not the inhabitants of Qumran, stored the scrolls. The skeletons beneath the rubble in the Cave of Letters may represent a population later than the one that stored scrolls closer to Qumran, and whatever they were storing in the cave may prove to be quite different.

According to Magen Broshi (2004), "many of the caves at Qumran and the surrounding area collapsed in antiquity, thereby covering early layers of occupation." In many of the caves closest to Qumran itself, including several man-made ones that had been dug into the soft marl terraces below the site, the entrances collapsed completely "in antiquity," sealing the caves. If any of this collapse occurred in the 31 BC earthquake, it seems logical that much of the scroll material may have been buried not by the attacking Romans or Jewish refugees, or even by Essenes seeking to hide them from the Romans, but by the 31 BC earthquake. How, then, would the inhabitants who survived the earthquake have retrieved their belongings? It may be that many of the artifacts and scrolls were not abandoned intentionally.

SCROLLS IN OTHER COLLAPSED CAVES?

Several caves in the area of Qumran remain unexcavated, since the roofs collapsed so completely that the form of the cave could not be determined. One such collapse may have blocked the entrance to a cave known to Dead Sea Scroll researchers as cave 4, causing it to be overlooked by archaeologists despite its proximity to Qumran (Magness 2002). Although the cave's entrance is clearly visible today from Qumran, archaeologists did not discover it in their early surveys of the area. Instead, it was first explored and plundered by Bedouins, who enlarged its entrance, dug another, and left thousands of scroll fragments hopelessly scrambled on its floor. I feel certain that undiscovered scrolls and other materials remain buried in other caves. These may be the only ones archeologists will be able to excavate before the Bedouin treasure seekers get there first and disturb the stratigraphy. In my view, archaeologists' best chance to discover more definitively what happened at Qumran, and how the Qumran settlement relates to the scroll materials, is to identify some of these sealed caves and excavate them systematically, before they can be looted by antiquities seekers.

In 1998, I helped launch a project to systematically survey desert caves throughout the Judean desert, not only the caves of Qumran. The Cave Survey program, a collaborative project between the Israel Cave Research Center at Hebrew University, the Land of Israel Studies Department at Bar-Ilan University, and Stanford University, focused on both the archaeological and geophysical aspects of the Judean desert caves.

We headed again to the ruins of Qumran. In the initial phase of this project, two hundred caves were identified, most of them tiny. Between October 2002 and August 2003, about fifty of these new caves, with locations between Ein Gedi in the south and Qumran in the north, were examined. We found that several of the deeper caves showed signs of collapse, possibly caused by tectonic movements. The date of collapse is still to be determined.

The project co-leader, Professor Amos Frumkin (2003) from Hebrew University, produced a more detailed description the major findings encountered in the caves during the survey. For example, in Cave #178, a small, remote cliff cave near En Gedi, the survey yielded artifacts dating from the Bar Kokhba period in the second century. In that cave, the artifacts discovered were remains of cloth, leather, animal bone, food, rope, baskets, early Roman glass pieces, and pottery. Also found were eleven coins from the Bar-Kokhba period (three coins are clearly marked with the word "Shimon," referring to Shimon Bar-Kosiba), a metal-tipped staff, twelve iron arrowheads of a distinctive style used by both the Romans and Bar-Kokhba's army, and even pieces of the wooden arrow shafts. Two folded papyrus scrolls written in Greek were found as well and forwarded to the Israeli Museum laboratories. Thus far, nothing has been found in the caves dating to the time of the Dead Sea Scrolls or Qumran's habitation periods. The artifacts were all found without actually excavating any of the caves, so what might remain buried deeper remains unknown.

Although only one of the caves in the survey has so far yielded artifacts of archaeological interest, the survey itself represents an unusual scientific collaboration between two academic disciplines—geophysics and archaeology—in which researchers traditionally have found it difficult to combine their insights toward a common goal. In this case, a whole new wave of archaeological exploration was initiated by an understanding of what earthquakes do. My hope is that, over the next decades, more attention will be paid to possible collapsed caves around Qumran. Perhaps we will find that earthquakes, in destroying the caves that sheltered the inhabitants of Qumran, may have preserved the past for future archaeologists. These places, undisturbed since their destruction by earthquakes, may provide the means to unravel the complicated and emotionally charged story of the Dead Sea Scrolls.

CHAPTER 7

Expanding the Earthquake Record in the Holy Land

Do not disturb yourselves at the quaking of inanimate creatures, nor do you imagine that this earthquake is a sign of another calamity; for such affectations of the elements are according to the course of nature, nor does it import anything further to men, than what mischief it does immediately of itself.

—Josephus, *The Jewish War,* a speech attributed to Herod the Great after the 31 BC earthquake in Judea

From collapsed caves full of artifacts near the Dead Sea to the abandoned mounds of tells in the desert, the remnants of civilizations past are literally everywhere in the Holy Land. In most places, though, the remains of the past are not sealed in caves for posterity but are constantly disturbed and displaced by the activities of modern life. As a child growing up near Haifa, I used to walk behind my father and his two mules as they plowed our field every spring. I remember picking up shards of ancient glass, pieces of simple mosaics and little squares of pink or white limestone exposed in the turned earth. Occasionally we would find a heavily oxidized copper coin or a few glass beads.

Our field was an inexhaustible source of these ancient pieces. On a few rare occasions, the plow was damaged when it struck a massive object beneath the surface; we would then dig up rough blocks of stone about 1 meter tall and 25 cm square in cross-section. We assumed that they were Roman mile markers of some kind.

I often wonder what ancient edifices used to stand where my father's field is today. I now know that the field lies in the path of the coastal branch of the *Via Maris*, the ancient Roman highway connecting the Mediterranean to the East. From our village, the *Via Maris* headed straight southeast to Megiddo, about 40 kilometers away, where it merged with the other northern branch, which comes southwest from Mesopotamia and Damascus through Hazor and Bet Shean. From Megiddo, the road continued south, parallel to the Mediterranean coast, all the way to Egypt. Thus, although my village is now something of a backwater town, it was once on the major thoroughfare of the ancient world. However, the specific source of the mosaic bits and ancient glass that our plow overturned remains a mystery, one repeated, I am sure, in fields throughout Israel.

Despite this continual disturbance of the ancient soil, however, a long habitation history and active seismicity, along with a dry, preserving environment, make the Holy Land one of the best places on earth to search for evidence of ancient earthquakes, and thus to extend the earthquake record into prehistory. This is important, since we have found evidence of ancient earthquakes in areas where there is no historical record of large, modern quakes. Israel, Jordan, Syria, and Lebanon, in particular, seem to be seismically much quieter today than in the past, leading to unrealistically rosy seismic hazard projections. N. N. Ambraseys, who may be the world's foremost expert on historical earthquakes, together with his colleague J. A. Jackson, compiled a list of known surface faulting in the eastern Mediterranean from historical sources, both in the "pre-instrumental" period (before 1894, when seismographs first began recording earthquakes in the area) and in modern times. They found that fault maps from the two periods look very different, with many sites of significant ancient faulting (including Israel) having almost no activity recorded by modern seismologists (Ambraseys and Jackson 1998).

Similar discrepancies show up in geological investigations, such as the excavation in Jordan at the site of the Karameh Dam. The dam, built across a wadi that emptied into the Jordan River, has the capacity to store fifty-five million cubic meters of water. One of its abutments sits astride the Jordan Valley Fault, part of the Dead Sea Transform. Sediments exposed during dam construction revealed deformation features from very large earthquakes that occurred there in the past, earthquakes as large as magnitude 7.8, with deformations more than twice the maximum allowed by the dam design (Malkawi and Alawneh 2000; Al-Homoud 2000). No earthquake that large has occurred at that site in modern times, and because of the region's low population density, together with its insignificance in biblical history, there is little historical material about ancient earthquakes for the area.

Clearly, archaeological evidence from nearby sites would be valuable here to help establish recurrence rates for these largest events. Despite the hazard, however, the dam was built as planned, and was completed in 1997. Jordanian officials must be ready for the inevitable failure of this project when the next big earthquake does occur.

As pointed out in earlier chapters, earthquakes are not one-time phenomena but rather the result of long-term seismic risk factors. To assess seismic risk throughout Israel, it seems that we would need a record longer than our modern instruments provide. Although this is true throughout the Mediterranean, I have focused much of my attention on Israel and its surrounding regions because of my heritage and connections there. This region, it seems, has a more complete historical record of earthquake damage than many other seismically active places, probably because of the region's worldwide religious significance. When towns that are mentioned in the Bible were damaged by an earthquake, it was a newsworthy event worldwide. Thus, many places that might have been considered insignificant economically or culturally were included in lists of affected towns because of their familiarity from the Bible stories. Likewise, descriptions of the damage in these biblical towns would be copied and disseminated far more broadly than for less notorious places, with the result that many of these reports, often second- or third-hand, survive to this day. Consequently, we can often draw fairly detailed maps of shaking intensity, even for earthquakes that occurred more than a thousand years ago.

This chapter describes several sites in Israel and Jordan that have been destroyed many times by earthquakes, and, for various reasons, rebuilt repeatedly. This is true of the majority of archaeological excavations in the region, largely because the mounds of accumulated debris, the tells described in chapter 2, alert archaeologists to the presence of artifacts beneath the surface. Doubtless there were many other ancient towns and isolated buildings abandoned after shorter periods that are less visible at the surface today, including the one that must have once stood in my father's field. It is at the tells where we find most of the data, though. The successive layers of destruction in these places, although they facilitate establishing a chronology for the buried layers, can make it difficult to interpret the causes of destruction because of disturbance from later activity.

MEGIDDO, JERICHO, AND JERUSALEM

A modest mound southeast of the Carmel Ridge in Israel, the tell of Megiddo rises 50 meters above the surrounding Jezreel Plain and covers some 6 hectares (15 acres) of land (Figure 7.1). This tell was first identified as the legendary Armageddon [the Greek corruption of the Hebrew "Har Megiddo" or Mount Megiddo] by a Jewish writer of the fourteenth century, Esthori Haparchi. It then was rediscovered by the British army officer H. H. Kitchener five hundred years later. Extensive excavations were conducted at the site by C. Fisher, P. Guy, and G. Loud (Yadin 1975). These archaeological studies revealed physical evidence of the historical development of Megiddo.

Megiddo's strategic importance, belied by its unimposing appearance today, stems from its unique topography (Figure 7.2). The land between the Mediterranean and the Jordan River served as a bridge between the civilizations in the South, in Egypt and Arabia, and those in the North, in Syria, Mesopotamia, and Anatolia. It

Figure 7.1 Aerial view of ruins of the ancient city of Megiddo (Armageddon).

was also a continuation of sea routes from the Mediterranean and the Gulf of Suez. The ruggedness of the region, however, crossed by several ranges of mountains and hills as well as the lowest valley on Earth, limited the possible routes for overland shipping or wheeled travel. The Carmel-Gilboa mountain range was a particular obstacle, and traffic from the Mediterranean to Syria and Jordan was funneled through a few gaps in the range. In fact, both the passes and the mountain range that obstructed traffic were products of tectonic motion along the seismically active Carmel-Gilboa fault system, which branches off from the Dead Sea Transform (see Figure 4.5).

The mound of Megiddo stands guard over one of the most important of these mountain passes, the Nahal Iron Pass, a traffic bottleneck on the main route between Egypt and Syria. Until the advent of more elaborate road construction by the Roman Empire in the first and second centuries AD, the gap at Megiddo was the only one that permitted the passage of chariots, though it was not an easy passage. A description from the latter part of the thirteenth

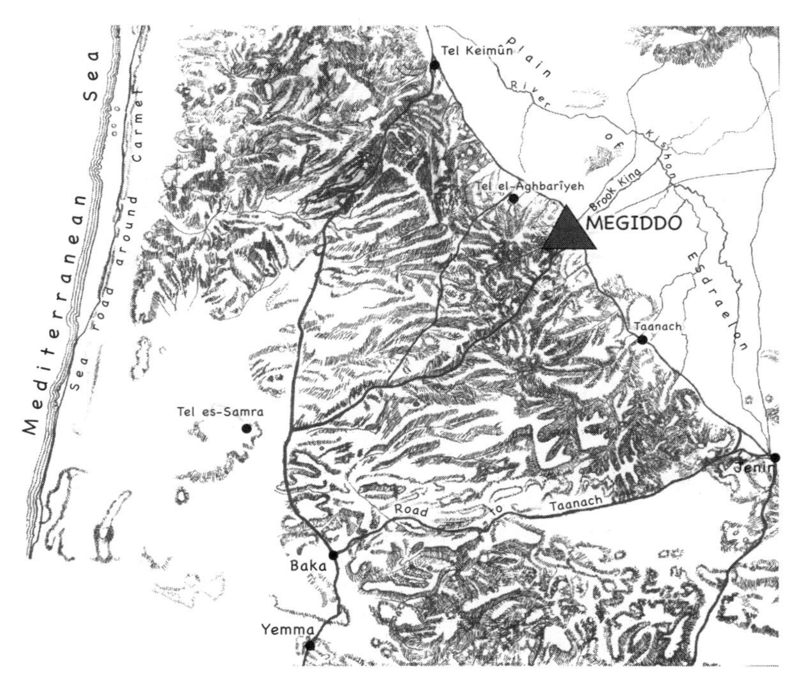

Figure 7.2 Location map of Megiddo showing its strategic placement in the local topography (after Nelson 1913).

century BC, found in the Egyptian Papyrus Anastasi I, gives some idea of the difficulty of the route:

> Thy path is filled with boulders and pebbles, without a toe-hold for passing by, overgrown with reeds, thorns, brambles, and wolf's paw. The ravine is on one side of thee, and the mountain rises on the other. Thou goest on jolting, with thy chariot on its side, afraid to press thy horse too hard. . . . The horse is played out by the time thou findest night quarters. (Hori [Egyptian Royal Official], Papyrus Anastasi I)

This narrow and difficult route made the pass from Megiddo particularly easy to guard, so whoever held power in Megiddo controlled not only the course of trade in the Fertile Crescent but that of war as well. Thus, the site figured prominently in some of the greatest ancient battles fought in this region. As Pharaoh

Thutmose III expressed it, "The capture of Megiddo is the capture of 1,000 towns." Indeed, fortifications were built and rebuilt there for close to five millennia, until around 500 BC (Finkelstein and Ussishkin 1994).

Four levels of destruction in the mound of Tel Megiddo are consistent with earthquake destruction, the lowest one attributed to the conquest by Thutmose III in 1468 BC. Why, however, would the Pharaoh have destroyed the place if his goal were to occupy the site and exact tribute? Although it is clear that Thutmose III conquered Megiddo, there is no more reason to assume that he ordered its physical destruction than to believe it was caused by an earthquake.

The second massive destruction at Megiddo, which occurred around 1250 BC, has variously been attributed to the Israelites or the Philistines, although historical evidence supports neither candidate. Again, however, the excavation of collapsed walls in Megiddo, and similar contemporaneous destruction in many nearby sites (Davies 1986), make the earthquake hypothesis a likely candidate.

The strongest evidence for earthquake destruction at Megiddo is probably the layer dating to between 1130 and 1000 BC, which some scholars attribute to conquest by King David's army. There is no historical mention, however, of David capturing Megiddo, much less leveling it, and, given the importance of the place at the time, it seems unlikely that such a conquest would go unheralded. More likely, Megiddo was destroyed by a massive earthquake, perhaps the same one that, according to the Bible, occurred during the battle of Michmash (1 Samuel 14:15). In the 1930s, excavators found collapsed walls from this period and, under the walls, smashed jars and several human skeletons, including the one shown in Figure 7.3. This layer is possibly contemporaneous with destruction layers from Dor, Gezer, and at least a dozen other sites in Israel and Jordan, all of them consistent with earthquake destruction.

The fourth layer of destruction occurred sometime after the conquest of Megiddo by Pharaoh Sheshonq in 925 BC. Although this layer is sometimes attributed to him, most documentary and

Figure 7.3 Victim of an earthquake. Excavations at Megiddo in 1932–34 uncovered a skeleton buried beneath the rubble of collapsed walls and smashed jars at Stratum VI. Note smashed jar (pithos) next to the body. The earthquake believed to be responsible is that of ca. 1020 BC, also implicated in the death of Doreen at Dor, 40 km west of Megiddo.

archaeological evidence indicates that Sheshonq did not destroy the city but rather had a monument erected in his honor there and exacted tribute from the residents. Yet, on the other hand, no definitive evidence has yet been excavated to indicate that an earthquake caused this destruction. We know that the massive earthquake during the reign of King Uzziah, around 760 BC, was important in this area—so important, in fact, that it is mentioned in the prophesies in the book of Zechariah (14:4–5):

> And his feet shall stand in that day upon the mount of Olives, which is before Jerusalem on the east, and the mount of Olives shall cleave in the midst thereof toward the east and toward the west, and there shall be a very great valley; and half of the mountain shall remove toward the north, and half of it toward the south. And ye shall flee to the valley

> of the mountains . . . yea, ye shall flee, like as ye fled from before the earthquake in the days of Uzziah king of Judah.

Other sources confirm that the earthquake Zechariah cites occurred around 760 BC. In any case, clearly it was a major enough disaster to be used as a temporal reference in Zechariah's prophecy. In fact, the prophecy includes such vivid details that it may actually describe the ground motion that occurred in the 760 BC earthquake, perhaps motion across a strike-slip fault.

Another example of an earthquake prophecy appears in the book of Revelation, where a mighty earthquake is prophesied to occur during the battle of Armageddon:

> And they assembled them at the place that in Hebrew is called Armageddon . . . and there came . . . a violent earthquake, such as had not occurred since people were upon the earth, so violent was that earthquake: And the great city was split into three parts, and the cities of the nations fell. . . . And every island fled away, and no mountains were to be found.

This account and the one in Zechariah may be examples of retrospective prophecies, a common feature in ancient literature where the author dramatizes a historic event as a kind of warning after the fact. This may indicate that, in the past, earthquakes were known to have struck Megiddo. Most important is that excavators keep in mind in future excavations that the area is subject to severe earthquake hazards, and always consider earthquakes as a possible cause for unexplained destruction there.

Topography is what made Megiddo so important militarily for so long, but other factors also determined the location of cities and forts in the ancient Mediterranean region. Another consideration in this arid land was the availability of water. The previous chapter discussed the importance of water in Qumran: dependent on cisterns to capture seasonal rainfall, Qumran was particularly vulnerable to earthquake damage. Cisterns were essential in many Israeli regions, and the collection, storage, and distribution of water was important throughout Israel's history. Any place with a year-round supply of fresh water, particularly clean, filtered water from a dependable spring, was prime real estate. One example was Jericho, described in chapter 3.

Tel Jericho's extraordinary stack of at least twenty-two layers of destruction has already been noted. With no modern city built on top of the tell—modern Jericho is a few miles from the remnants of its ancient namesake—archaeological excavation has progressed relatively unhindered, despite the usual impediment of political unrest in the area. This has been a liability in some ways, since excavations using early archaeological methods clearly destroyed at least as much information as they revealed. Still, because of these excavations, the repeated destruction and rebuilding of Jericho has been acknowledged for some time. With its long record and repeated rebuilding, Jericho would be a great candidate for a sort of key site, a place where evidence of repeated earthquake damage could be correlated with destruction in neighboring sites where the record might be less continuous. If we could systematically examine the written evidence of earthquakes and try to tie together the archaeological stratigraphy among the sites scattered around Israel and Jordan, we might be able to piece together a much more informative record than we could hope to gain from any one site alone. A great deal of archaeological work has been done to try to correlate the remains at Jericho with other sites throughout the Holy Land. Kathleen Kenyon (1978) was particularly comprehensive in her synthesis of the archaeology of the region. Although she does mention the general likelihood of earthquakes in this region and specifically ascribes certain damage layers to earthquakes, she makes no explicit effort to determine the range of such earthquake damage; she does not mention whether the earthquake in question could have damaged nearby towns.

One of the most suggestive layers in Jericho is one Kenyon dated, based on pottery styles, to between 1600 and 1550 BC, where the walls of Jericho collapsed, burying storage jars full of grain (Figure 7.4). The grain was carbonized by a fire concurrent with the collapse, and carbon dates of the grain have placed it remarkably close to Kenyon's original estimate (Bruins and van der Plicht 1996).

Figure 7.4 Jars full of carbonized grain found beneath a collapsed wall in Jericho, dated to ca. 1600–1550 BC by Kathleen Kenyon (from Kenyon 1953).

According to Kenyon (1979, 177–178), many towns in the region suffered total destruction at the same time (though she does not implicate an earthquake), including the sites of Tel Beit Mirsim, Hazor, Shechem, and Megiddo. Among these sites, only Megiddo was immediately rebuilt, with no apparent change in culture; most of the others were abandoned for more than a century.

In chapter 3, I speculated that an earthquake may have figured prominently in the biblical story of Joshua's destruction of Jericho. Does the archaeological record preserve any evidence at all of Joshua's Jericho? One of the early expeditions reported the discovery of a city wall from Joshua's time, but Kenyon (1957) indicated that the dating of that wall was incorrect. As described in chapter 3, she found only one small remnant of Jericho that could have coincided with the traditional estimate of 1400 to 1250 BC for Joshua's conquest of the site. If a larger habitation did exist at the site, the remains of that culture have all but vanished, eroded away during a long period of abandonment.

Recently, however, archaeologists have reexamined both the archaeological site at Jericho and radiocarbon dates for other events in the ancient world. One particularly notable theory links the plague of darkness in Egypt, before the Exodus, to the explosive eruption of the island of Thera in the Aegean Sea, an event dated to the end of the seventeenth century BC. If this date is correct, the nearly missing layer Kenyon attributed to Joshua's time would be far too recent, and the destruction from ca. 1600 would be a more likely candidate.

The debate over the assignment of this layer, based on samples of locally made pottery unearthed at Jericho, and on radiocarbon dates from the charred grain and from bits of timber found in the destruction layer, has at times been heated, almost vicious. (For an example of two diametrically opposed interpretations of the same evidence, presented side-by-side in one issue of *Biblical Archaeology Review*, complete with *ad hominem* attacks, see Bienkowski 1990 and Wood 1990.) It will be some time before the dust settles on this issue. For now, the archaeology is the source of more questions than answers, and until some of those questions can be resolved, using Jericho as a key site will remain difficult.

Although the mound of ancient Megiddo was eventually abandoned, and the modern city of Jericho has abandoned the old tell, the city of Jerusalem has been continuously inhabited for at least four thousand years (Cline 2004). The reason for its continued importance reaches beyond favorable topography or water supply to that least tangible reason for the persistence of cities: religious significance. King David's choice of Jerusalem as the capital city of his Hebrew nation, and, probably more important, his decision to move the Ark of the Covenant into the city, sealed its fate as an enduring human habitation into modern times.

Despite its long history, most of the city's archaeological secrets are nearly impossible to reveal. The whole city has been densely inhabited for millennia, with almost no abandonment in any quarter. Modern and ancient homes, churches, and official buildings block access to the layers of rubble beneath them. When a scrap of real estate is not actively occupied, it is usually because it is imbued with some intense political or religious importance and digging into it would touch off a firestorm of controversy.

As a result, excavations carried out in Jerusalem are often quite limited and piecemeal, undertaken only when an opportunity happens to present itself. This occurred in 1948, for example, when, during the Arab-Israeli war, a mortar hit a building that, according to tradition, housed the tomb of King David. In the process of repairing the damage inflicted by the explosion, the archaeologist Jacob Pinkerfeld (1990) also did a cursory excavation of one portion of the structure. Beneath the floor, he found over half a meter of debris, including three earlier floors. In fact, further investigations have suggested that the walls of this building were part of the Church of the Apostles, built in the first century AD at the place where the Last Supper was thought to have occurred. The walls have been destroyed and rebuilt many times, but because of the site's importance to those who believe it was the tomb of David, the excavation was hindered and no further investigation was possible.

Still, although the city cannot be systematically excavated, modern instruments can probe beneath the surface to tell us not how old the layers are or what artifacts are in them but at least whether specific parts of the city are founded on archaeological debris or on solid ground. A government study by Israel's Geological Survey was released in 2004, confirming what archaeologists had long suspected: the Old City is mostly founded on the rubble of previous constructions. As discussed in chapter 2, this makes the Old City particularly vulnerable to earthquake damage.

The historical record bears this out. In nearly every historical report of earthquakes affecting Jerusalem, damage is reported either to the Temple, the Old City walls, or both. Of course, the importance of this site and its role as the focus of religious pilgrimages for three of the world's major religions guarantees a record that is unparalleled in any other region. We have written records from many sources for earthquakes in Jerusalem, most of which also affected large regions of the countryside.

A recent geological investigation of sediments from the Dead Sea (Ken-Tor et al. 2001), confirms the validity of many of the historical reports mentioned in this chapter. As described in chapter 2, when shaking of sea-floor sediments is severe enough, the loose, water-saturated sediments lose their strength and flow like a liquid. When the shaking stops, the sediments settle and resolidify, leaving behind a chaotic, mixed layer that is readily identified, a *seismite*. These layers frequently contain organic material that scientists can date by Carbon-14 dating.

The 2001 study examined layers from the Ze'elim Terrace on the shore of the Dead Sea, a stack of sediments exposed by rapid modern drops in Dead Sea water levels (caused by diversion of the fresh water sources that feed the Dead Sea). There are gaps in this sequence, marking drought periods where the ancient water level in the Dead Sea was lower than usual and there was no sediment deposited at the Ze'elim Terrace site. However, during the periods when sediments accumulated there, every major historical earthquake on the Dead Sea Transform has been correlated to a seismite. This is a great new resource and one that can be extended in the future. The periods for which the Ze'elim Terrace contains no data could be examined by drilling into the deeper sediments in the Dead Sea floor. Thus, we can confirm the historical record at Jerusalem, and use it, along with the archaeological record in more accessible sites nearby, to build a physical earthquake stratigraphy for the region.

AD 1927

The easiest way to discuss the earthquake history of this region is to work backward from the present day, focusing on some of the largest or historically most significant earthquakes. I begin with the most recent major earthquake to strike what was then known as Palestine, an earthquake briefly noted at the end of chapter 2. The July 11, 1927, Palestine earthquake—or the 1927 Jericho earthquake, as I will call it, since Jericho was so devastated by it—was recorded at more than one hundred seismological stations in Europe, South Africa, North America, and Russia (Ben-Menahem et al. 1976). Although these records were few, as the science of seismology was still young, they were sufficient to provide accurate arrival times for the waves generated by the earthquake. From these, seismologists tried to determine the epicenter of the quake, but confusion and unreliable reports probably led to an error in location. The epicenter was initially located next to the river Jordan, about 15 kilometers north of Jericho, under the plain that extends north of the Dead Sea. However, a more recent reevaluation (Shapira, Avni, and Nur 1993; Avni et al. 2002) of the primary evidence from the earthquake indicates that a more reasonable location would be on the western shore of the Dead Sea itself (see Figure 4.5) and that the original placement of the epicenter was biased by reports of severe damage in the more northerly area. Some of this damage was based on unreliable, secondhand reports, and the rest of the damage probably occurred because this northern part of the fault did slip during the earthquake, even though the starting point, the epicenter, was farther south.

This earthquake was the very first damaging earthquake to be seismically recorded in the region, and at magnitude 6.2 (Shapira 1979; Ben-Menahem et al. 1976), it remains to this day the largest earthquake to shake the Dead Sea region since the proliferation of modern seismographs. More recently, many small earthquakes have been recorded there, especially since networks of seismic recording stations were placed in Egypt, Israel, Jordan, Lebanon, and Syria (Figure 7.5). The trenching data from the Karameh dam show that the area may be subject to much larger earthquakes.

The damage in the 1927 earthquake was widespread. Many towns and villages experienced shaking of intensity VII or greater, including, of course, Jericho, where there was almost total destruction, and Nablus, far to the north, about which the Jewish Telegraphic Agency reported, on July 14, 1927, "with the exception of two streets, the entire city lies in ruins." There were also accounts of severe damage in Ramleh, Lydda, and Jerusalem (Amiran, Arieh, and Turcotte 1994; Avni et al. 2002), and cracks in the Old City wall, cited in the 2004 Israeli government survey as hazardous, probably date from that earthquake. Although the quake was only

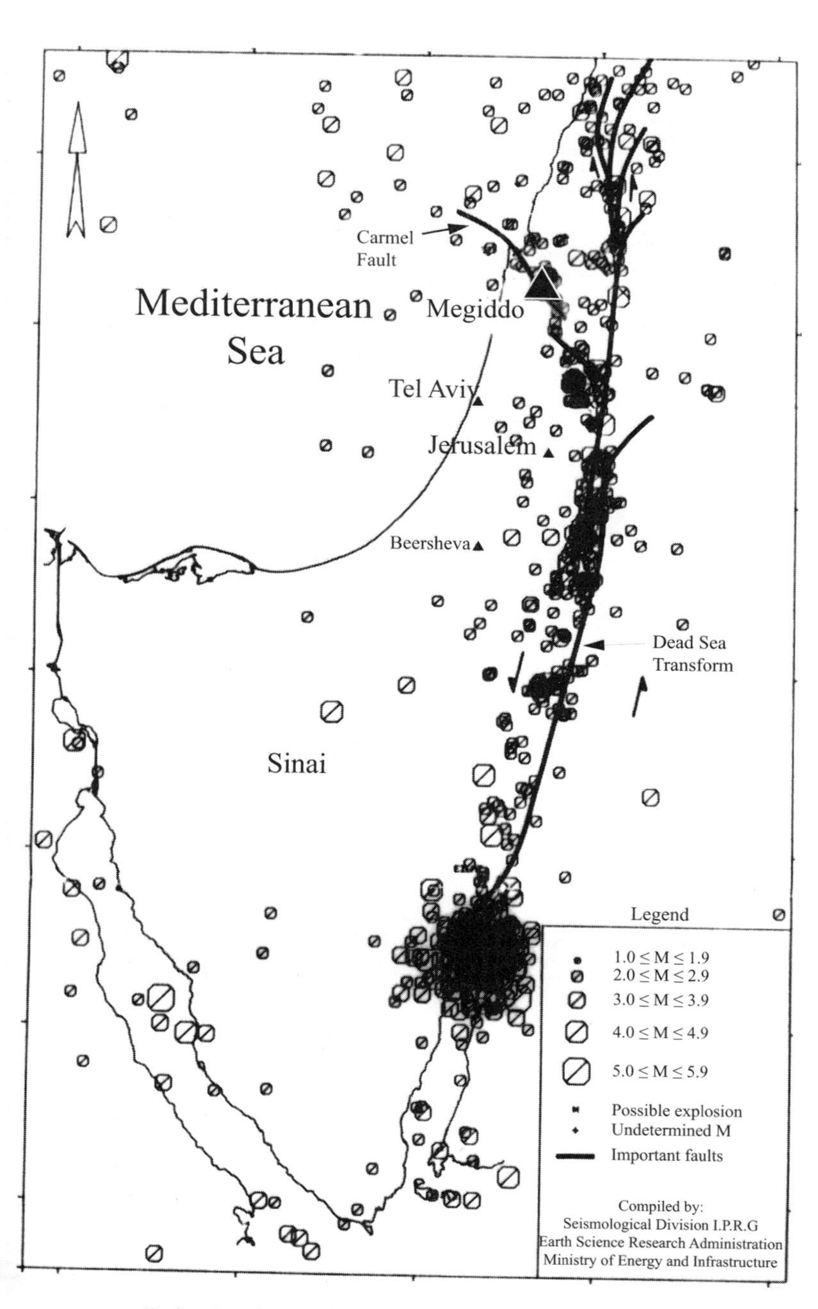

Earthquake epicenters in and around Israel during 1981-1988

Figure 7.5 By recording the accumulation of small earthquakes on a map of the Dead Sea region, we can delineate the fault systems in the region.

Figure 7.6 Bystanders in Nablus sifting through the rubble of collapsed buildings after the 1927 Jericho earthquake.

moderate in magnitude, the damage to structures extended from the Negev in the south to the coastal plain, lower Galilee, and the Sea of Galilee in the north, a distance of 100 kilometers or so. Among other areas affected in this earthquake were As Salt and Karak in Trans-Jordan, Amman, Irbid, Nazareth, Ramallah, Reine, Hebron, Jerash, Tiberias, Raula on the coastal plain, and Tel-Aviv. In total, 285 people were killed, and about 1,000 were injured. An earthquake of greater magnitude would have caused destruction over a much larger area. Figure 7.6 shows the deep fissures and soil cracks that formed in the ground around the countryside.

The following accounts of the devastation caused by this earthquake appeared in the *Los Angeles Times* (1927):

> The Allenby Bridge was damaged at both ends. Other bridges are down. The Greek Catholic Church of the Holy Sepulchre has been declared unsafe because of the cracks in the walls. The roof of the chemical laboratory of the Hebrew University on the Mount of Olives collapsed

> . . . The Government House, located on the Mount of Olives, was badly damaged. The private rooms of the High Commissioner, Lord Plumer, and Lady Plumer, who are now on vacation in England, were wrecked. This building was also declared unsafe by the authorities. (Jewish Telegraphic Agency, July 12, 1927)

> Two synagogues, one in Jerusalem, the other in Tiberias, were destroyed. In several Palestinian towns the Moslem mosques and the Government office buildings were damaged. The house of the British representative in Amman, Transjordania, was totally destroyed. The Church of the Holy Sepulchre, as well as the Greek Choir Chapel and two large domes are damaged. (Jewish Telegraphic Agency, Jerusalem, July 13, 1927)

> A terrible picture of destruction is to be seen at Nablus, the biblical city of Shechem, seat of the ancient Samaritan sect and home of a large Arab population. With the exception of two streets, the entire city lies in ruins. Rescue parties are at work extricating the bodies of victims from among the ruins in Nablus and neighboring village. (Jewish Telegraphic Agency, Jerusalem, July 14, 1927)

This 1927 earthquake stands as a bridge between modern instrumental earthquake seismology and the history of ancient earthquakes. On the one hand, seismographs around the world recorded the shaking for this event, and scientists in many countries analyzed and shared the data they collected. On the other hand, the construction methods in most of the affected countryside were primitive, with structural designs virtually unchanged for millennia. There had never been a major earthquake here since the advent of modern seismic engineering, so there had been no attempt to consider earthquake safety in building designs:

> The primitive wall throughout the East was a mud wall. It is still the common resource. Galore pretentious structures are faced with stone and the facing often presents the appearance of blocks accurately dressed to smooth bearing surfaces, but it is a sham. The edges on which the blocks adjoin are but an eighth to a quarter of an inch wide; they rest on that narrow edge; behind it they are chipped to a rude pyramidal form, which is stuck into the mud or rubbish. Walls of this kind are from two

> to three feet thick or more. In the old, closely built-up sections of an ancient city, a Jerusalem or a Nablus, they are piled up with the most extraordinary disregard for support. (Willis 1928)

Health and sanitation were similarly poor, and emergency relief resources and methods were pitifully inadequate to handle the disease, shortages, and general chaos in the aftermath of the earthquake. With cracked cisterns drained of water, desperate people drank whatever water they could find, which resulted in widespread death from cholera and dysentery. Wind-driven clouds of dust from the heaps of dry rubble caused irritated, infected eyes as well as respiratory problems. Had such an earthquake occurred thousands of years earlier, one can well imagine that the aftereffects would surely not have been much different.

The impact of the earthquake in Jerusalem remains visible today. Steel bands, I-beams, and concrete reinforcements still hold many of the damaged buildings together (Figure 7.7). Oddly enough, the city's current inhabitants remain largely unaware of this earthquake. My aunt, Hana, who moved to Jerusalem in 1938, thought I was joking when I told her that an earthquake had devastated the city only some ten years before she arrived there. She is not alone; most people in Jerusalem scoff at concerns of earthquake safety.

The question of seismic hazard, of course, remains quite relevant today. Efforts to modernize construction methods can have only a limited benefit in a city where many of the buildings are hundreds of years old. Even distant earthquakes can have tragic consequences. The results of the relatively small 1927 quake help us imagine how past earthquakes along the Dead Sea plate boundary—some of which would have been of much greater magnitude—must have caused widespread devastation in this region. This sort of regional destruction where dozens of villages are affected, thousands of houses collapse, and thousands of lives are lost—all within only the few minutes it takes seismic waves to traverse the area—is one of the hallmarks of earthquake destruction. The aftermath of such an event resembles a huge war zone, covering thousands of square kilometers.

Figure 7.7 Interior of the Church of the Holy Sepulchre, showing the steel beams that were installed to repair the damage caused by the 1927 earthquake.

The series of aftershocks that followed the 1927 Jericho earthquake were strong enough that scientists were able to locate their epicenters. These aftershocks cluster along a line that extends northward from the relocated source, suggesting that the fault runs approximately from north to south within the trough of the Jordan River (Freund et al. 1970; Garfunkel 1981; Nur and Reches 1979; Salamon et al. 1991). (Again, this northward propagation of the earthquake along the fault explains why Jericho was so badly damaged, even though it was not that close to the epicenter.) Analysis of the seismic records using modern methods has confirmed this hypothesis and specified further details of the motion across the fault: the land mass on the east side moved northward, whereas the land mass on the west moved southward during the earthquakes, with a total offset of as much as 50 cm. The results fit other information obtained from geologists and geophysicists (Garfunkel et al. 1981; Joffe and Garfunkel 1987).

AD 1546

At 1:00 PM on January 14, 1546, a large earthquake struck the Holy Land, badly damaging Jerusalem. Ambraseys and Karcz (1992) quote an anonymous chronicler of the event:

> It lasted a short while and calmed down, and generally, there was not a tall house in Jerusalem that was not left destroyed or fissured, and the same in [Hebron]. In Gaza, the madrasa of Qayitbey was destroyed as well as the south part of his madrasa in Jerusalem, and its north and east sides; also, the top of the minaret over the Bab as-Slisila was destroyed. In Nablus, the earthquake was stronger than elsewhere, and 500 lives were lost under the ruins.

Other towns where damage was reported included Safed, Tiberias, Ramla, Hebron, Gaza, As Salt, and Karak (Ken-Tor et al. 2001). This is a relatively large area, more than 120 miles from north to south, and reaching from the Mediterranean coast to the territory east of the Jordan River, for which the historical record is not as complete. It is difficult to determine where the epicenter of this earthquake might have been, but the area of damage apparently centered on the Dead Sea Transform. There were seismic sea waves induced in both the Mediterranean (from Gaza to Tel Aviv) and in the Dead Sea, so that is not much help. Both these seismic sea waves could have been triggered by landslides or by movement of the sea floor caused by the earthquake's slip.

According to the anonymous source, damage was severe in Jerusalem:

> There is no house that was not destroyed or cracked, and even from the new city wall there fell a scythe in height, such as at the Gate of Mercy. And also fell the Ismalite mosques as well as the cupola of [al Aqsa], and so did the [Holy Sepulchre]. . . . And the gentiles say that there never was such an earthquake in Jerusalem . . . and in contrast, praise be to God, our synagogue was left undamaged. About 12 Ismalites perished, and none of the Jews. But in Nablus about 560 Ismalites perished of the townfolk, but nobody knows of the villagers, since they still may be buried under the rubble. (Ambraseys and Karcz 1992)

This account, almost certainly exaggerated, is typical of the reports made by religious residents of Jerusalem throughout antiquity; every earthquake was an opportunity to find evidence that God favored one people (or synagogue) over another.

The damage to the Church of the Holy Sepulchre, however, is well documented. It is particularly interesting to compare historical drawings of the church, three of which are shown in Figure 7.8. The first drawing is the oldest and possibly depicts the church tower as it was originally constructed. The second, made after the 1546 earthquake, shows the damage described in the quote above; because the government forbade rebuilding, a new roof was simply added to what remained of the original. The tower endured further damage in the 1927 earthquake, and its current appearance is depicted in the photograph where it is another increment shorter and again re-roofed. Other structures in Jerusalem show similar signs of repeated earthquake damage, and their religious significance has ensured that the remains are preserved to this day.

AD 749

A short time after the end of Byzantine control of the Holy Land and the conquest by the Seleucid Arab armies, a strong earthquake completed the destruction of what the war had spared. Previously damaged buildings were destroyed, and some building projects begun by the King of Damascus were abandoned after the earthquake. In Khirbet Mina, near the Sea of Galilee, archaeologists in the 1950s uncovered partial walls, built to a height of a meter or so. Outside the walls, and in the spaces that were to become rooms, stacks of worked limestone blocks were neatly arranged in rows, ready to be used. Another site, the Hisham Palace complex outside Jericho, which had been under sluggish construction for some twenty years, was similarly abandoned after the quake. Construction never resumed, partly because the economy had been weakened by war, and partly because the economic power that remained had shifted away from the Mediterranean in the West to the eastern Arab kingdom in Damascus.

Figure 7.8 The changing face of the Church of the Holy Sepulchre in three stages: (a) woodcut from the fifteenth century; (b) drawing by Dominik de la Greche after the 1546 earthquake (black arrow shows where the top of the tower is missing); and (c) photograph taken after the 1927 Jericho earthquake.

The historical record for this earthquake reports damage to some six hundred towns and cities throughout what are now Israel and Jordan. Jericho, Bet Shean, and Susita, discussed in other chapters, were among them. The earthquake devastated Bethlehem, where one of the few structures left standing was the Church of the Nativity (International Millennium Publications [IMP] 1999). In the cities of Pella and Jerash in today's Jordan, the earthquake leveled many structures, trapping skeletons as described in chapter 5. It spawned a tsunami on Israel's Mediterranean coast and seiches in both the Sea of Galilee and the Dead Sea. This earthquake was a final blow to the region after the Arab conquest, and almost nothing was rebuilt. The conquest had produced economic weakness, making it impossible for the economy to withstand the physical blow of the earthquake.

From a few ancient writings and many archaeological excavation logs, we can extract evidence for concurrent destruction around AD 749, giving us a general idea of the geographic extent of the damage from this earthquake. The region affected was slightly larger than that affected by the 1927 earthquake, and so we can expect that the magnitude was also somewhat greater, perhaps 6.7 to 7.2.

Some sources also mention a large earthquake in AD 747, but others believe the two earthquakes were the same. This illustrates one of the most difficult problems in correlating archaeology to the historical record. Dates sometimes differ by only a year or two, and sometimes it is clear that an error was probably made by one source or another. It complicates matters that several different calendars were used at various times in the region, depending on which group had power at the time.

Usually, archaeology is of little help in distinguishing between one large earthquake and two smaller ones only a few years apart. Even in an ideal situation, where coins from the year of the earthquake were buried under the rubble in two separate earthquakes, there is no way to know whether the earthquake happened the year the coins were minted or years later. After all, most of us carry around coins minted decades ago, and it would not be unusual at any given time not to be carrying a current year's coin.

Unfortunately, when this earthquake (or two of them) occurred, the water level in the Dead Sea was at an unusually low level, so that the now exposed sediments examined in the seismite study I mentioned earlier (Ken-Tor 2001) were also above the water line. With no geological record of this earthquake in those sediments, we have not yet seen the kind of chaos such a large earthquake would leave in the Dead Sea sedimentary record. Seismites from this earthquake must exist, of course, since the reported seiches in all of the region's water bodies testify to the violence with which the sea-floor sediments were shaken. To find them, however, geologists would have to sample deeper-water sediments by drilling and retrieving cores from the Dead Sea floor.

AD 363

One of the best-documented earthquakes from the first millennium occurred in AD 363. Evidence for this earthquake has mounted over the years, emerging from historical documents and archaeological digs. The most definitive historical record is a letter discovered by Sebastian P. Brock in the Harvard Library, which he later translated and published as *Harvard Syriac 99* (Brock 1977). The letter, attributed to Cyril, the bishop of Jerusalem from AD 350 to 388 (Russell 1980; Browning 1989), records the effects of a great regional earthquake.

The letter describes the damage resulting from an earthquake occurring "on Monday at the third hour, and partly at the ninth hour of the night . . . on 19 *Iyyar* of the year 674 of the Kingdom of Alexander the Greek," which corresponds to Monday, May 19, 363. This precise dating helped to settle confusion between this earthquake and another that occurred in 365, probably centered on the island of Crete and causing tsunamis throughout the eastern Mediterranean. The 363 earthquake was of particular social importance, because it interrupted a project of great religious significance to both Jews and Christians: the rebuilding of the Jewish temple in Jerusalem on the site of the Second Temple, which had been destroyed by the Romans in AD 70:

> At the digging of the foundations of Jerusalem, which had been ruined because of the killing of the Lord, the land shook considerably, and there were great tremors in the towns round about. . . . We have not written to you at length, beyond the earthquake that took place at God's (behest). For many Christians too living in these regions, as well as the majority of the Jews, perished at that scourge—and not just in the earthquake, but also as a result of fire. . . . At the outset, when they wanted to lay the foundations of the Temple on the Sunday previous to the earthquake, there were strong winds and storms. . . . It was on that very night that the great earthquake occurred, and we were all in the church of the Confessors, engaged in prayer . . . and the entire people . . . drove out the demons of the city, and the Jews, and the whole city received the sign of baptism, Jews as well as many pagans. (Brock 1977)

Harvard Syriac 99 lists twenty-three cities that were badly damaged in the 363 quake, shown on the map in Figure 7.9. Although Jericho is not specifically mentioned, it almost certainly suffered some damage in this quake, considering its proximity to other badly damaged sites.

The archaeology of the area also indicates that many other cities were damaged. Some of the best archaeological evidence for this event was found in excavations at the ancient city of Petra, described in chapters 4 and 5. Household objects and coins found under collapsed structures were interpreted as earthquake damage as early as 1955, and more recent excavations continue to support that interpretation. Of special importance in the dating of the AD 363 earthquake was a purse full of coins, found with two skeletons under the rubble of one of the rooms. Ian Browning (1989) described the find:

> It is significant that the initial analysis of this hoard shows that there are coins dating up to, but none after, AD 363. It is reasonable, therefore, to assume that the destruction of this house was caused by the same earthquake of AD 363 which wrought such terrible damage in Petra generally.

Earlier coin evidence had been found in 1976 and 1977, when archaeologists in Petra excavated the "middle house" destroyed by an earthquake:

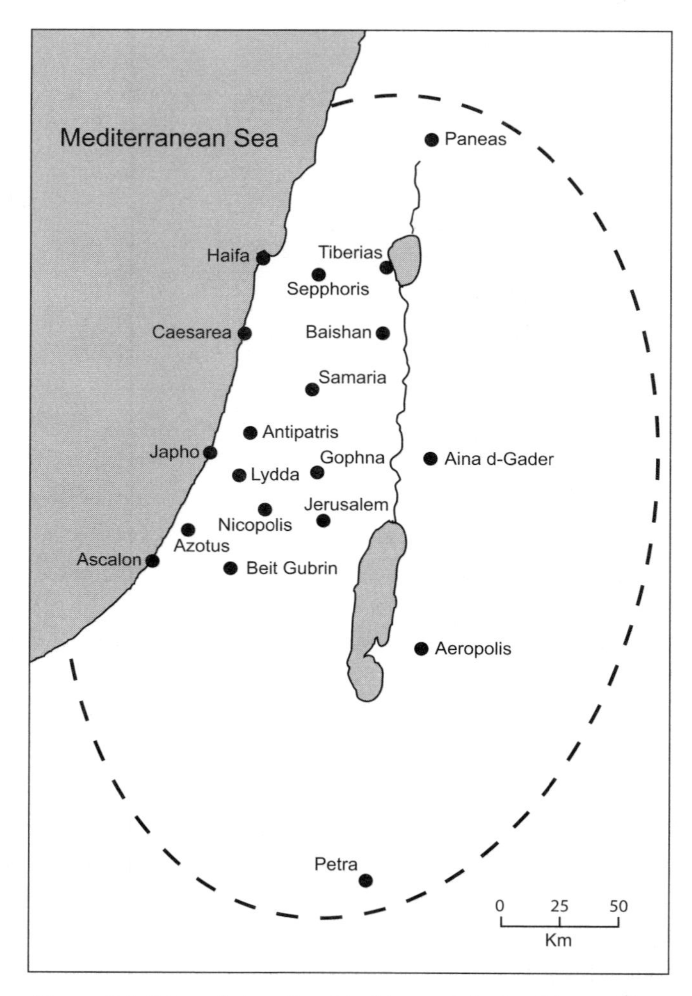

Figure 7.9 Map of sites listed in Harvard Syriac 99 as suffering destruction during the earthquake of 363. The dotted line indicates the estimated felt area based on historical evidence (after Guidoboni et al. 1994, *Catalogue of Ancient Earthquakes in the Mediterranean Area up to the 10th Century*).

Near the doorway, and close to where the end of this cupboard would have been, there was found a hoard of 85 small denomination bronze coins. These lay amongst the crushed remains of three ceramic vessels, any or all of which may have originally contained them. The 45 identifiable coins from this hoard were all minted during the reign of Constantius II (337–361). (Russell 1980)

Most of the identifiable coins were minted after a currency reform in 354 AD. We therefore know that the earthquake could not have happened before the bronze coinage reform, and that the coins were still in use at the time the structure buckled. This is consistent with the 363 AD date.

The author of *Harvard Syriac 99* cited various miraculous signs that supposedly accompanied the earthquake as further evidence of God's wrath. Although the signs he described have not been corroborated with physical evidence, the earthquake itself has. This physical corroboration, along with the very specific list of towns (many of little biblical significance), in which the letter even specifies the extent of destruction for each town, make this source, despite its clear religious bias, an important chronicle of the damage area.

The archaeological evidence from Jerusalem is much more sparse than that from Petra because of the difficulty of excavating in and around a living city. As noted earlier in this chapter, Jerusalem is particularly hard to excavate, because so many of its areas have religious significance for one group or another. Still, according to *Harvard Syriac 99*, more than half of Jerusalem was destroyed in the earthquake, so evidence of the destruction should be practically everywhere should excavations be undertaken. Indeed, one excavated site near the Temple Mount revealed a collapsed, burned domestic structure, with coins dating to the appropriate time. Whether this house was destroyed by an earthquake is not certain, but several sources believe it is likely (Brock 1976; Russell 1980).

The damage in Jerusalem from this quake probably had a more lasting effect on local and world culture than the damage anywhere else. The project to rebuild the Jewish Temple of Jerusalem on the site of the Second Temple had been championed and funded by the pagan Roman emperor Julian the Apostate, less out of sympathy with the Jews than from animosity toward Christians, who opposed the Temple reconstruction. Christians of the time saw the 363 earthquake as a sign from God, chastising the emperor and proving to the Jews that the Temple would never be rebuilt. To them, this was an extension of Jesus' prophesy of the Temple's destruction, recorded in Matthew 24:1–2.

Fourth-century Jewish writings are mysteriously silent about both the Temple project and the 363 AD earthquake, a strange omission, as both subjects would seem to have been of primary interest to the Jewish community at the time. All surviving descriptions of the events surrounding the earthquake come either from the pagan Romans or from Christian historians, most of whom did not witness the events as they occurred. Newman (1842) provides a good summary of accounts from the century following the event.

The tales are rather wild, including a whirlwind, thunder, and lightning preceding the quake, and even fire falling from the sky and melting the workers' tools. Then, following the quake, accounts describe fireballs chasing the workers around the neighborhood as they tried to flee, as well as spectral crosses, appearing both in the sky and impressed upon the bodies of the laborers.

It is difficult to extract much of historical value from Newman's sources, but the archaeological evidence makes it clear that at least the earthquake happened. Further archaeological work may well uncover evidence of fire, as so often happens following earthquakes. Widespread fires after the quake might be the original source of the fantastic fireball descriptions. All sources agree, however, that the earthquake occurred during Julian's attempt to restore the Temple and just before his death on June 26, 363, a date that agrees with the coin evidence found in Petra. Evidence of this earthquake is preserved as a thin seismite in the Dead Sea sediments at the Ze'elim Terrace (Ken-Tor et al. 2001).

JUDEA, 31 BC

Unlike in the earthquake of 363 AD, for which *Harvard Syriac 99* provided precise details, history has not provided an itemized list of the damage from the 31 BC earthquake. Nevertheless, Flavius Josephus (1991b [AD 93], bk. 15, chap. 5) did leave one account:

> At this time it was that the fight happened at Actium, between Octavius Caesar and Antony, in the seventh year of the reign of Herod; and then it was also that there was an earthquake in Judea, such a one as had not happened at any other time, and which earthquake brought a great destruction upon the cattle in that country. About ten thousand men also perished by the fall of houses; but the army, which lodged in the field, received no damage by this sad accident.

In addition to this information, archaeological excavations from parts of the Holy Land other than Jerusalem uncovered further evidence for the 31 BC quake. The most famous site destroyed by this quake, of course, was Qumran, as described in chapter 6, but Qumran was not the only city affected. Geological trenching across the Jericho Fault revealed that this event caused fault rupture at the surface near Jericho itself (Reches and Hoexter 1981), only 15 kilometers north of Qumran (recall that Jericho is where Herod's winter palace suffered severe damage). The Ze'elim Terrace study found widespread seismites from this quake, which is not surprising since the part of the Jericho Fault that ruptured is only 60 kilometers from the Ze'elim Terrace.

Reports of the calamity in Judea quickly reached the Arabs in Jordan (Josephus 1991a [AD 75], bk. 1, chap. 20). As so often happens with word-of-mouth reports immediately following a disaster, the accounts exaggerated both the damage and loss of life:

> In the meantime, the fame of this earthquake elevated the Arabians to greater courage, and this by augmenting it to a fabulous height, as is constantly the case in melancholy accidents, and pretending that all Judea was overthrown. Upon this supposal, therefore, that they could easily get a land that was destitute of inhabitants into their power, they first sacrificed those ambassadors who were come to them from the Jews, and then marched into Judea immediately.

Survivors of the earthquake were demoralized, thinking that God had abandoned their cause. Indeed, the army and the populace were ready to declare defeat. To rally his people, Herod delivered a speech. Most of the address dwelt on the evils of the Arabs and the victories the Jewish army had so far enjoyed on the battlefield, but, for us, the most remarkable element of the speech was the short, almost offhand mention of the earthquake. Although I quoted Herod's address above, in chapter 3, this ancient recognition of the nature of earthquakes bears repeating:

> And do not disturb yourselves at the quaking of inanimate creatures, nor do not imagine that this earthquake is a sign of another calamity; for such affections of the elements are according to the course of nature; nor does it import anything further to men, than what mischief it does immediately of itself. Perhaps, there may come some short sign beforehand in the case of pestilences, and famines, and earthquakes; but these calamities themselves have their force limited by themselves, without foreboding any other calamity. (Josephus 1991a [AD 75], bk. 1, chap. 20, p. 4)

Heartened, the Jewish army mounted a defense, and Herod managed to lead his army to victory.

The Jordan River Valley is most notable for the almost continuous record of military conflict there, but it is also a seismically active region. As the earthquake of 31 BC showed, it was inevitable that a military conflict and an earthquake would eventually coincide, perhaps with great, historical consequences. In 31 BC, Herod's local forces managed to overcome the physical and psychological damage that the earthquake inflicted. The locals in Jericho, in Joshua's time, might not have been so fortunate.

JERICHO, IN THE TIME OF JOSHUA

Our survey of historical earthquakes in the region around Jericho ends in 31 BC. Earlier than that date, the historical record is spotty, and the only written source we have for most events is the Bible. The earthquake that occurred around 1050 BC was discussed in chapter 3, in conjunction with the battle at Michmash, for which archaeologists have found, at best, possible evidence in Megiddo and Dor. One of the earliest biblical accounts, which does not explicitly mention earthquakes but has remarkable parallels to some of the historical damage, is the account of Joshua's attack on Jericho. Was an earthquake responsible for the biblical fall of Jericho?

The Bible, as noted earlier, is not an entirely reliable source of historical information, particularly about the figure of Joshua, and, as discussed in chapter 3, controversies plague the archaeology at Jericho. If the written record is so unreliable, and archaeology is equivocal, what makes us think there is something in the Bible worth examining? Could we not dismiss this account as mere fiction? Probably not, considering what we have learned from biblical archaeology. Oral traditions, though perhaps unreliable in their details, often have a basis in fact. Archaeologists have repeatedly found evidence for events once widely thought to be mythological. It makes sense, therefore, to ask whether logically explainable events could have given rise to the tale of Joshua's legendary feat. That an earthquake could have struck during the battle at Jericho is certainly plausible; recall that the 31 BC earthquake struck during the battle at Actium and the 1050 BC earthquake during the battle at Michmash.

For now, therefore, I examine what the Bible says about Jericho's conquest, and see how it could fit with the seismicity of this area and the archaeological evidence from the tell. The biblical story clearly parallels a few of the stranger aspects of modern earthquake damage in this area, and some of the events that seem unrelated miracles could be consequences of a single earthquake. To examine this possibility, we must look at the whole story from the Bible. Of particular importance is the recorded miracle of the Israelites' passage into Canaan in the Book of Joshua (3:14–7):

> And it came to pass, when the people removed from their tents, to pass over Jordan, and the priests bearing the ark of the covenant before the people; and as they that bare the ark were come unto Jordan, and the feet of the priests that bare the ark were dipped in the brim of the water (for Jordan overfloweth all his banks all the time of the harvest) that the waters which came down from above stood and rose up upon a heap

> very far from the city Adam, that is beside Zaretan; and those that came down toward the sea from the plain, even the salt sea, failed, and were cut off: and the people passed over right against Jericho. And the priests that bare the ark of the covenant of the Lord stood firm on dry ground in the midst of Jordan, and all the Israelites passed over on dry ground, until all the people were passed clean over Jordan.

How does this passage have anything to do with an earthquake? The key lies in the geology and topography of the Jordan River.

In many places, including the region the passage alludes to, the banks of the Jordan River are very steep and composed of soft, unstable sediments. Thus, when an earthquake occurs, the riverbanks often collapse into the river, partially or completely obstructing the flow. Recall that this effect was mentioned in the letter from the Jerusalem resident after the 1546 quake, and it has been historically documented many times. During the 1927 Jericho earthquake, chunks of mud slid into the river Jordan near Damiya (referred to as "ancient Adam" in Joshua), about 40 kilometers north of Jericho, temporarily reducing its flow (Ben-Menahem et al. 1976). Earlier, in 1906, an earthquake had caused the flow to be interrupted for twenty-four hours. Other recorded disruptions, typically lasting one or two days, include those of 1834, 1534, 1267, and 1160. Thus, the combination of the destruction of Jericho and the stoppage of the Jordan is typical of earthquakes in this region.

If this account indeed refers to an earthquake, the chronology of events has probably been altered. Many biblical apologists appeal to multiple earthquakes to produce the effects necessary to match the biblical account. In other words, one earthquake dammed the Jordan so that Joshua could cross, and then another earthquake, a few days later, collapsed the walls while Joshua was marching around the city. This is not impossible; major earthquakes usually do have aftershocks that continue for weeks or months after the main event. As discussed earlier, however, there is no hope that an archeological excavation could determine whether this happened. Even given perfect preservation of the earthquake rubble, there is no way to distinguish between one earthquake and two in rapid succession. Nor do I believe it is necessary to invoke two separate earthquakes.

The inconsistencies elsewhere in the book of Joshua make it clear that the author, or authors, of the accounts therein, or the oral historians who passed the stories down until they were recorded, were not above embellishing for dramatic effect. If a single earthquake struck, both damming the river and destroying Jericho's walls, Joshua would have arrived to find that the city had already been conquered for him, apparently by God. Surely, even that circumstance would have seemed like miraculous timing to Joshua's approaching army, and would have cemented in their minds that they had a divine right to the land they were invading. The embellishments of marching around the city seven times, the shout, and the sound of trumpets would all make the story more dramatic in the retelling as it was passed down through the years.

Another feature of the biblical story was that the conquest happened after the harvest, when the grain had been gathered into the city. This would be consistent with the grain found under the city walls, and would be typical of an earthquake. Since grain was very valuable in ancient times, if the city had simply been conquered without the walls collapsing, the conquering people would certainly have taken the grain. On the other hand, if the people of Jericho merely left for some reason, and the walls collapsed later, they would surely have taken the grain with them. Alternatively, if there had been a prolonged siege before the walls fell, the grain would have been consumed by the besieged people.

Evidence of an earthquake is also indicated by the discovery of the skeletons of two people killed by Jericho's fallen walls (Garstang and Garstang 1940). In addition to these individuals having possibly been earthquake victims themselves, the skeletons also bear witness to the continual seismic activity in this region. The description of the second male is striking:

> A man's head was found to have been completely severed from his body, as may be seen in [Figure 7.10]; but as the excavation continued we noticed a continuous fissure across the floor of the room and

Figure 7.10 Two skeletons of people crushed during an earthquake at Jericho (ca. 1400 BC). Note the apparent decapitation of the skeleton on the right, caused by fault motion long after the victim's death (From Garstang and Garstang 1940).

> running up the walls, telling of an earthquake which by a remarkable coincidence had subsequently produced this curious illusion of decapitation. (Garstang and Garstang 1940, 51)

Obviously, this fissure in the floor and walls was caused by an earthquake; but, just as clearly, it could not have been the same quake that killed the person whose skeleton was so curiously distorted. The earthquake that caused the decapitating fault motion must have occurred long after the one that collapsed the building, at a time when the remains of the earlier earthquake victim had been buried by debris and reduced to a skeleton. Because these remains are so old, and because this is such a seismically active region, the bones have no doubt been shaken many times since they were pinned beneath the fallen wall.

Figure 7.11 A view of the many layers of Jericho's walls unearthed during the 1953 excavation under the direction of Kathleen Kenyon.

As the illustration in Figure 7.11 testifies, Jericho's walls preserve clear evidence of repeated destruction. In the words of Kenyon and Tushingham (1953, 865):

> Repeatedly we saw places where a wall had been built 5,000 years ago, only to fall, be patched, then rebuilt, then strengthened in front and behind, and finally, when repairs proved fruitless, be replaced by another broader wall on top of its ruins. Not only in our day has security lain in preparedness! . . . Jericho's walls were repaired or completely rebuilt no fewer than 16 times! The earliest wall was undoubtedly destroyed by an earthquake; we found it lying flat, fallen forward on its face. Later walls probably suffered the same fate. Others may have been destroyed by enemies as the 17 certainly was. The destruction of this last wall marks a great catastrophe for Bronze Age Jericho, as indeed it must have for the whole of Palestine. Its predecessor had collapsed, possibly because of an earthquake.

Figure 7.12 is a close-up photo of the layers of ash and rubble that make up the Jericho archaeological site.

Figure 7.12 A closer view of some of the layers from Jericho. The dark layers are evidence of repeated, site-wide conflagration.

If indeed an earthquake was responsible for the Israelites' capture of Jericho, it profoundly affected the political future of the region. That it was considered not an earthquake but an act of God may be just another intriguing and peculiar example of the human response to earthquake destruction. The enormous psychological impact of an earthquake on the survivors is not surprising, of course. The sudden shaking of the earth beneath our feet shatters an assumption on which we base our daily actions: that we can always rely on the solidity of the ground beneath our feet. When terra firma is suddenly not so firm, we are stunned.

We see traces of this repeatedly in the historical record. Every major earthquake is "such a one as had not happened at any other time" (Josephus 1991b, bk. 15, chap. 5), and in every quake the survivors see the hand of God. In his insistence that earthquakes were simply natural phenomena, Herod was indeed a visionary. Nevertheless, as they have throughout the centuries, earthquakes will continue to shape culture and politics in the Middle East.

CHAPTER 8

Earthquake Storms and the Catastrophic End of the Bronze Age

The whole world quaked, the slopes of Ida with all her springs
and all her peaks and the walls of Troy and all Achaea's ships.
—Homer, *The Iliad*

This book began with a visit to the ruined city of Mycenae in Greece, which was destroyed at the end of the Bronze Age, around 1200 BC. The hypothesis that it was destroyed by attacking Sea Peoples came about largely because archaeology had been unable to explain the destruction there, especially in the context of apparently contemporaneous destruction at other sites throughout the eastern Mediterranean region. This catastrophic destruction at the end of the Bronze Age is still a source of controversy among archaeologists, historians, and other scientists. Although at many of the sites in question, archaeologists have independently proposed earthquake destruction, the archaeological community has largely dismissed the idea that earthquakes played a large part in the wider catastrophe.

Robert Drews (1993) compiled a map showing the sites that have been found in the eastern Mediterranean that were destroyed at the end of the Bronze Age (Figure 8.1). Drews names forty-seven Aegean and eastern Mediterranean sites where archaeological evidence suggests some level of catastrophic collapse between 1225 BC and 1175 BC, a period of about fifty years. Despite the

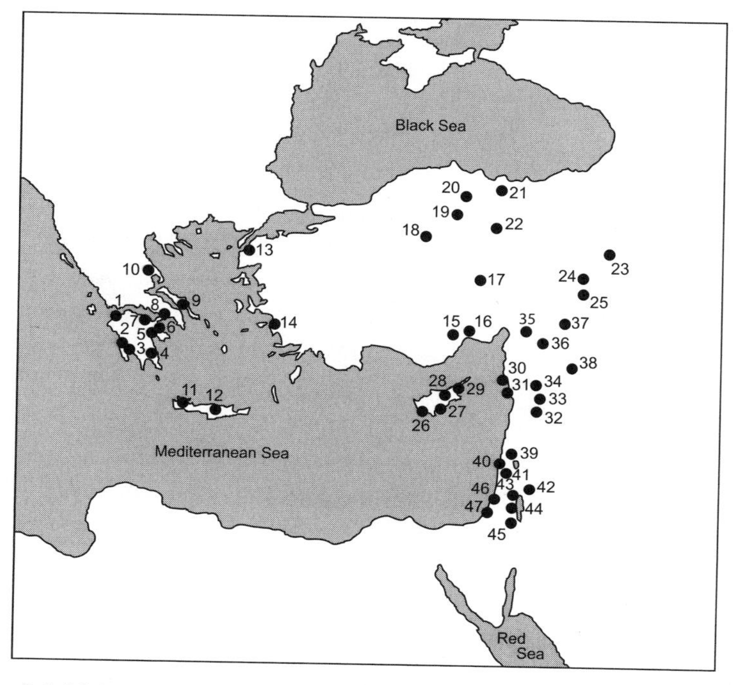

Figure 8.1 Major sites in the Aegean and Mediterranean that were destroyed ca. 1225–1175 BC at the end of the Late Bronze Age (after Drews, 1993). Destruction is probable but not certain at sites given in italics.

1. Teichos Dymaion
2. Pylos
3. Nichoria
4. The Menalaion
5. Tiryns
6. Midea
7. Mycenae
8. Thebes
9. Lefkandi
10. Iolkos
11. Kydonia
12. *Knossos*
13. Troy
14. *Miletus*
15. Mersin
16. Tarsus
17. Fraktin
18. Karaoglan
19. Hatushash
20. Alaca Höyük
21. Masat
22. Alishar Höyük
23. Norsuntepe
24. Tille Höyük
25. Lidar Höyük
26. Palaeokastro
27. Kition
28. Sinda
29. Enkomi
30. Ugarit
31. Tell Sukas
32. Kadesh
33. Qatna
34. Hamath
35. Alalakh
36. Aleppo
37. *Carchemish*
38. Emar
39. Hazor
40. Akko
41. Megiddo
42. Deir 'Alla
43. Bethel
44. Beth Shemesh
45. Lachish
46. Ashdod
47. Ashkelon

ongoing debate about what caused the collapse, most scholars agree that severe destruction was widespread. Drews (1993) notes, "Within a period of forty or fifty years at the end of the thirteenth and beginning of the twelfth century almost every significant city or palace in the eastern Mediterranean world was destroyed, many of them never to be occupied again." What caused the collapse?

The previous chapters make it clear that the archeological record is peppered with the remains of earthquake damage, and that based on modern seismicity much more must still remain unidentified. The frequency-magnitude relation explained in chapter 2—the fixed relationship between the number of small, frequent earthquakes versus large, infrequent earthquakes on any given fault—ensures that destructive earthquakes must have happened many times in antiquity, in most of the major cities of the ancient world. This is as close to an indisputable fact as we can have in earth science.

Earth science can show that earthquakes must have happened in a given area or at a given site. It can even show that specific earthquakes did happen, using geological evidence including ancient fault ruptures, tsunami deposits, and soil liquefaction. However, geology alone cannot correlate these lines of evidence with periods of human habitation. The long years of dedicated work by archeologists who establish chronology by matching pottery, written records, coins, and other datable finds are invaluable in determining which civilizations were affected by these earthquakes, how large the affected areas were, and how the societies responded. Unfortunately, the bias of archaeologists against earthquakes interferes with efforts to assemble the kind of body of earthquake evidence that could be most useful, to both earth scientists and archaeologists themselves.

This conflict becomes especially apparent in excavations dealing with the complex issues surrounding the end of the Bronze Age. If archaeologists' interpretations of many earthquake-damaged layers from this period, at sites scattered around the Mediterranean, are taken as a group, they constitute perhaps the greatest catastrophic collapse in the human experience. The most persistent argument against the earthquake hypothesis is the sheer size of the quake required to cause such damage or the unlikely coincidence of so many earthquakes occurring in such quick succession. I address those concerns later in this chapter. But given this skepticism about how widespread the effects of an earthquake could be, some archaeologists seem determined to prove that earthquakes could not even have happened during the time in question.

One of the most cogent alternative explanations for Bronze Age destruction is offered by Robert Drews (1993), who proposes that a revolution in the techniques of warfare were the main cause for the upheaval at this time. He argues convincingly that this revolution indeed took place, and that certainly it must have unsettled the ancient world considerably. What is odd, however, is his implication that this revolution somehow rules out an earthquake catastrophe in the same region at the same time.

Drews entirely dismisses the idea of earthquakes, systematically dismantling the arguments of archaeologists who proposed earthquake destruction at their own excavations, including Blegen et al. (1953, 1958) at Troy, Mylonas at Mycenae, Schaeffer at Ugarit, Samson at Thebes, Evans at Knossos, Kilian at Tiryns, and Davis and Kempinski at Megiddo. Schaeffer, Kilian, and Blegen, in particular, also extended their hypotheses to include regional destruction of some extent.

Drews, in his refutation of their hypotheses, makes some interesting arguments. He dismisses Schaeffer's and Kilian's proposals that Alalakh, Hatushash, and Pylos were destroyed by earthquakes, with the justification that the original excavators of those sites disagreed, and that archaeologists generally accept the original excavators' conclusions regarding the fate of a given site. However, in cases where those same investigators—or others—posited earthquake destruction at other sites they excavated, Drews proceeds to refute their findings systematically. Arguments he uses to favor human destruction over earthquakes include the presence of foreign weapons; the absence of skeletal remains, which is simply wrong for most of the places he cites; the presence of widespread fire, which he naïvely says is unheard of in ancient earthquakes; and the absence of ground displacement in the excavated site, another example of naïveté. He even uses arguments that the original excavators offer as evidence favoring earthquake destruction: where there is evidence of tilted or collapsed walls, he favors subsidence unrelated to earthquakes; where there is evidence of fire, he assumes it was deliberately set; where there is clear evidence of skeletal remains trapped within rubble layers—as at Mycenae, Tiryns, Midea, and Knossos, information that came to light well before his own book was published—he simply says that it does not exist.

With all his negation of earthquake evidence, Drews implies that the success of his own interpretation of events, that a revolution in techniques of war was responsible for the collapse, depends on proving not only that earthquakes had nothing to do with the collapse but also that quakes could not have occurred at the same time as a revolution in war technology. The cause is either war or earthquakes—if one, then not the other. Nor is he alone in this mind-set. Archaeologists who favor theories of agricultural collapse, anonymous aggressors, climate change, or meteor strikes also tend to argue against the possibility of earthquakes as exacerbating agents. Although any one of these extraordinary events could explain some subsets of the archaeological or geological data from that era, the fact is that earthquakes, even large ones, are *not* extraordinary events in this region; they are an ever-present threat there, and evidence that they occurred in no way negates other theories. The either/or debates that rage in archaeological circles imply a strange separation of the human world from the natural world, as if earthquakes would hold off until human conflicts were resolved, or as if political and military strategists would not alter their plans in the face of a sudden natural disaster. The suggestion that the two are inevitably linked is seen as a capitulation, a sign of a weak theory that must be bolstered by unlikely coincidences.

CLAUDE SCHAEFFER CITES EARTHQUAKES

The archaeologist Claude Schaeffer (1948; 1968, 753–768) first made the bold statement that earthquakes were responsible for much of the destruction. Schaeffer attributed the speedy and unanticipated ruin of all the Hittite cities in Asia Minor, in the early twelfth century BC, to earthquakes: "Our inquiry has demonstrated that these repeated crises which opened and closed the principal periods . . . were caused not by the action of man." Although Schaeffer was a great archaeologist, this idea got him into trouble with his peers, and, as I suggested in chapter 1, spawned an anti-earthquake bias among archaeologists and historians that persists to this day. For example, Drews argues:

> It is about as clear as such things can be that the cities destroyed in the Catastrophe were destroyed by human hand. Theories that an act of God destroyed many of the most important eastern Mediterranean cities *ca.* 1200 not only are unsupported by archaeological evidence but go against the historical evidence for both antiquity as a whole and for the period *ca.* 1200 specifically.

It is unclear exactly what Drews means when he contends that earthquake theories conflict with the historical evidence for antiquity as a whole, but it may merely be an example of the irrational anti-earthquake bias rearing its head again. A map of historical seismicity (Figure 8.2) shows earthquakes of magnitude 6.5 and larger (maximum intensity of VII and larger) in the Aegean and the eastern Mediterranean between 1900 and 1980 (Armijo, Deschamps, and Poirier 1986). The map is a snapshot of seismic activity over a period of approximately eighty years in Greece, Turkey, Syria, Israel, and Lebanon, and is based not on inferences from writings but on recordings from modern seismic instruments developed shortly after the turn of the twentieth century. The recordings tell us the locations and magnitude of these earthquakes, and the Richter magnitude scale lets us roughly calculate the maximum ground motion that can be expected anywhere in the surrounding region, even where there are no seismic recordings.

Figure 8.3 shows a simple sketch of the major tectonic plate boundaries responsible for seismicity in the region of the Bronze Age collapse. The Arabian plate is moving north relative to the African plate (which is also moving north, but more slowly), the

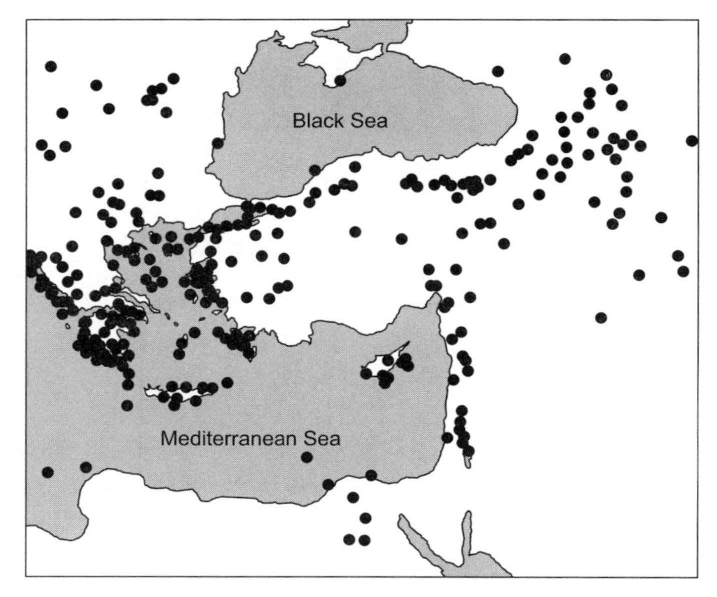

Figure 8.2 The Eastern Mediterranean: instrumentally recorded earthquakes of magnitude 6.5 and larger in the period ca. 1900–1980 (after Armijo, Deschamps, and Poirier 1986).

Anatolian plate is moving west, and the European plate is moving southeast. The tectonics of the Aegean region, in particular, are complicated, because here the end of the Anatolian plate fragments into several smaller "micro-plates" that splay out across mainland Greece from east to west (Jackson 1993), all moving west at various speeds. As a result, the Aegean region is a hotbed of seismic activity (Dewey and Sengör 1979; McKenzie 1970, 1972; Galanopoulos 1973; Rapp 1982).

Geologic evidence indicates that the configuration and dynamics of these faults have not changed appreciably for hundreds of thousands or even a few million years. Because the plates move only a few centimeters per year, the landscape and the plate-bounding fault patterns change slowly (Ambraseys 1970, 143). As discussed in earlier chapters, as far as human history and prehistory is concerned, the earthquake pattern of the twentieth century is a good representation of what has been experienced in this

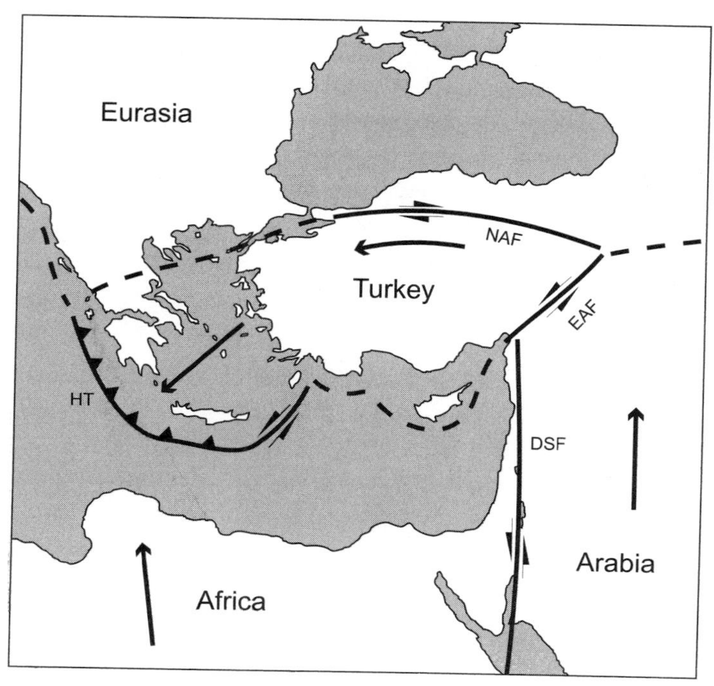

Figure 8.3 A simple sketch of the plates in Turkey and the Aegean, showing the relative motions between the Africa, Arabia, Turkey, and Eurasia plates, accommodated by strike-strip faulting on the North (NAF) and East (EAF) Anatolian Fault zones and by shortening in the Hellenic Trench (HT). Black arrows are the approximate directions of motion of Arabia, Turkey, the southern Aegean, and Africa relative to Eurasia. DSF is the Dead Sea Fault zone, C is Crete, and P is Peloponnese (after Jackson 1993).

region for as long as the human race has existed (cf. also Drews 1993, 37).

To determine what this ground motion means for human constructions, we need to know earthquake intensities in the area, measured, for example, on the modified Mercalli intensity scale (described in chapter 2). Figure 8.4 is a map of those intensities for the eighty-year period from 1900 to 1980. The gray-shaded regions are where ground shaking from earthquakes reached an intensity of VII or greater—the threshold for significant damage in modern circumstances—sometime over this period.

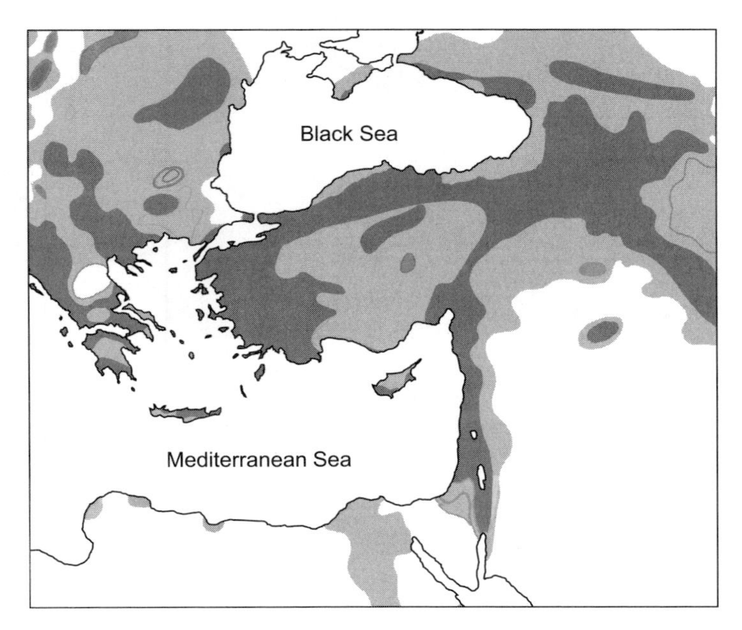

Figure 8.4 Map showing the maximum intensity of seismic ground motion in the Aegean and Eastern Mediterranean during 1900–1980 (modified from Karnik 1968). The maximum local intensity within the deepest brown regions during this period is greater than VII on the Modified Mercalli Scale.

The frequency-magnitude relation described in chapter 2 implies that the earthquakes on a given fault obey a simple statistical rule: because there are many more small earthquakes than large ones in a given time period, the average frequency of large ones can be predicted if we know the frequency of small ones. For example, Figure 8.5 shows a cumulative frequency function of earthquakes in Greece over a sixty-year period that had a magnitude of 4.0 or greater.

If calculations of frequencies were made for the entire Mediterranean, the figure would be similar to that given in Figure 8.5. The line would have the same slope (because the constant B is similar for the broader region), but because the area covered would be larger, the total number of quakes would be greater and the line would be shifted upward. Conversely, if calculations were made for a smaller area—say, only for Crete—the figure would have

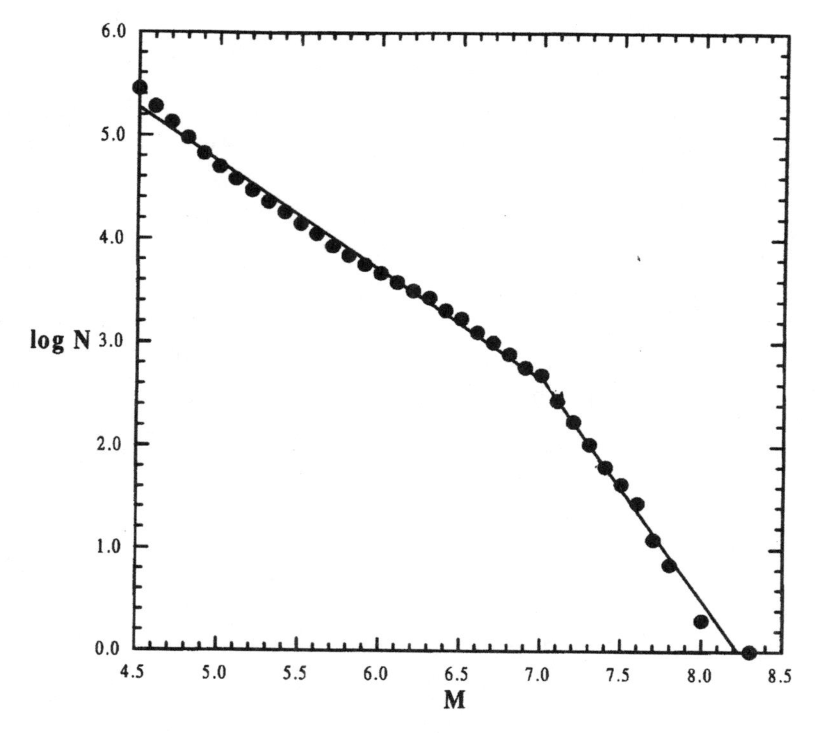

Figure 8.5 Frequency-magnitude plot of earthquakes in Greece during 1911–1969 as developed by Papazachos and Papazachou 1997. If a longer period of earthquakes were observed, we would not expect the shape of the lines to change.

roughly the same slope but with fewer total events. If we collected data for eight hundred years instead of eighty, we would again obtain the same curve, except that it would be shifted upward, because the time window would be ten times longer, and we could expect roughly ten times as many earthquakes. The longer the period one examines, the greater the chance of finding a *bigger* earthquake as well. Since very large earthquakes are much less frequent than small earthquakes, we must observe for a long time to have a good chance of seeing a large one.

It is important to remember, however, that there is an upper limit to earthquake magnitudes in any given region. Also in chapter 2, I discussed the relationship between the magnitude of an earthquake and the length of the fault that slipped during the event. We know, for example, that in Turkey and Greece there were earthquakes in the past exceeding magnitude 8.0; however, based on the lengths of faults in the area, the cutoff may be somewhere between 8.2 and 8.5.

The map in Figure 8.4 is a mildest-case estimate, based only on earthquakes we have already seen; we have no assurance that our eighty-year window detected the largest quakes expected for this region. If larger quakes occur, the gray regions will expand, and more of the region will be mapped inside the high-intensity contours. The point is that even a short, eighty-year catalogue reveals much information about the seismicity in an area that can be extrapolated over much longer periods of time.

These data cannot reveal, however, *when* the next big earthquakes will strike or when they struck before we began to keep records. The longer we wait inside the gray region, the more likely it is that we will be jolted by high-intensity shaking. The longer a civilization has existed, the greater the probability that it experienced a devastating earthquake. So if a city within the gray region of Figure 8.4 is several thousand years old, it is virtually inescapable that its buildings and facilities experienced very damaging earthquakes. We do not need archaeological records or historical accounts to reassure us of this statistical near-certainty. Nevertheless, archaeology is indispensable in determining the timing, because, as far as we know, there is no regular pattern to the occurrence of large earthquakes.

The forty-seven sites that, according to Drews (1993), were destroyed in the Bronze Age catastrophe are shown in Figure 8.6, superimposed over the intensity VII ground-motion map of Figure 8.4. Most of the sites fall within or close to the high-intensity regions, and must have been badly damaged by earthquakes at least once in the past. Indeed, Kilian (1996) noted that "the Mycenaean sites for which there is archaeological evidence of seismic activity coincide with areas that have been hit by earthquakes with a magnitude M=6.0 or larger during the last two hundred years."

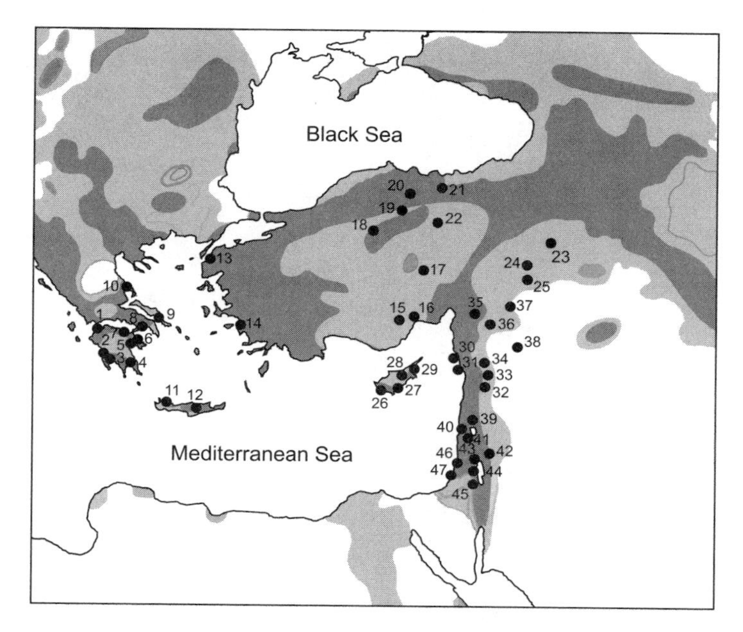

Figure 8.6 The Aegean and Eastern Mediterranean sites destroyed ca. 1225–1175 BC (after Drews 1993, Figure 8.1), superimposed on the maximum intensity of seismic ground motion in the Aegean and Eastern Mediterranean during ca. 1900–1980, in which the intensity was greater than VII (after Karnik 1968, Figure 8.4).

Of course, this does not prove that these sites were destroyed by earthquakes at the end of the Bronze Age, but it does place earthquakes high on the list of suspects whenever unexplained destruction is found. The excavators' reports for many of these sites describe damage that could have been caused by earthquakes as easily as by human hands. That, along with the known seismic risk throughout the region, makes it indefensible to dismiss earthquakes without serious consideration.

HATUSHASH AND TROY, AND THE ANATOLIAN FAULT

Two historic earthquakes illustrate the seismic risk in the important Bronze Age cities of Hatushash and Troy. Both sites are within the

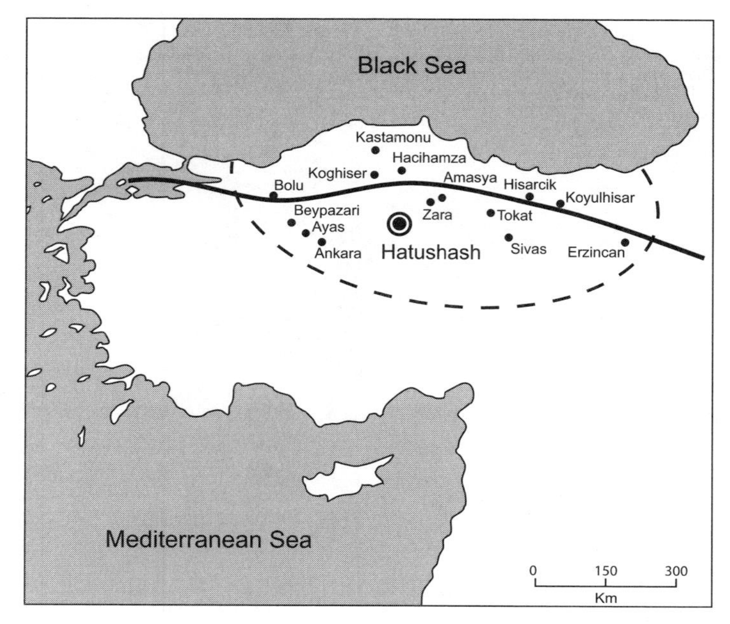

Figure 8.7 Location map of the Anatolian earthquake of August 17, 1668. Place names given in full are sites affected by this earthquake. The dashed ellipse shows the approximate intensity I > VII (after Ambraseys and Finkel 1988). Also added is the location of the ancient Hittite capital Hatushash, which falls well within the damage zone of the 1668 earthquake.

seismic hazard zone associated with the North Anatolian Fault, which runs across all of northern Turkey, then through the Dardanelles into the Aegean Sea and across Greece, and terminates east of Greece in the Hellenic Trench (see Figure 8.3).

The first of these historic earthquakes occurred on August 17, 1668, with a probable magnitude between 7.5 and 8.5. Ambraseys and Finkel (1988), who pioneered many of the concepts considered here, estimated and mapped the extent of the intensity VII region from reports of damage in villages and towns at the time (Figure 8.7). I have added to their map the location of Hatushash, the capital of the Hittite Empire from 1700 to 1200 BC. To a geophysicist, this map can mean only one thing: if an earthquake like the one in 1668 happened during the period when Hatushash

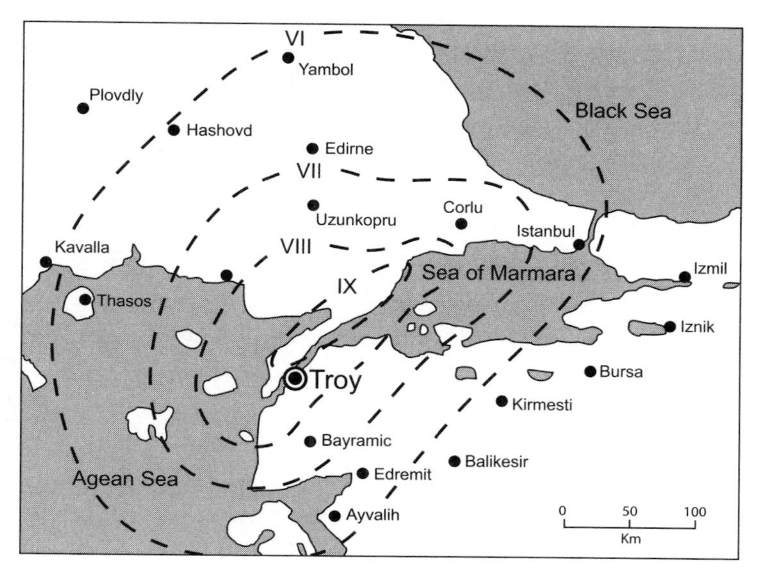

Figure 8.8 Isoseismal map of the Saros-Marmara earthquake of August 9, 1912, M = 7.4. Note the location of Bronze Age Troy in northwestern Turkey, well within the intensity VIII contour of this earthquake. Had this kind of earthquake occurred when Troy was inhabited, the damage would have been massive (after Ambraseys and Finkel 1987).

was important, that city could not have escaped significant damage considering the building methods used in those days. There is no historical record explaining the decline of Hatushash, but archaeology shows that it was destroyed suddenly around 1200 BC, and the entire city burned. Some have ascribed the destruction to the Sea Peoples.

The second example is an earthquake that struck in western Turkey on August 9, 1912 (Figure 8.8), with intensity contours again estimated from contemporary damage accounts. Once more I added to Ambraseys and Finkel's map, this time marking ancient Troy, which is inside not only the intensity VII area but also the intensity VIII area, which indicates heavy damage on the modern Mercalli scale. A quake of this size, though registering only intensity VIII today, might well have been a IX at Troy in ancient times, as ancient construction methods were poorer than today's.

Of course, Troy was only an archaeological site in 1912, but we can say with certainty that similar earthquakes visited this region in antiquity—probably every few hundred years, according to frequency-magnitude predictions. If this kind of earthquake struck while Troy was a thriving city, it could in no way have escaped severe and possibly total destruction. Blegen and his colleagues were justified in proposing earthquakes as the cause of some of the destruction they found there (Blegen 1963; Blegen et al. 1953, 1958).

An earthquake at Troy, then, is not at all a farfetched proposition, and an earthquake at Hatushash is also quite likely. Our seismicity hazard map, which, remember, is still an incomplete picture of the seismic risk in the area, indicates that an earthquake would be perfectly plausible in any of the cities on Drews's list of Late Bronze Age destruction. Blegen et al. (1953, 1958) and Schaeffer (1948), who suggested this notion years ago, could have had the right idea. Troy and Hatushash could not have escaped the violence of earthquakes. Why, then, was Schaeffer subjected to ridicule and Blegen to resistance from other archaeologists and historians?

One reason is that Schaeffer proposed a single, huge earthquake to explain the simultaneous destruction over the entire large region. Many at the time believed that such large earthquakes could never happen. Schaeffer's proposal was published well before the magnitude 9.5 Chilean earthquake of 1960, which is still the largest instrumentally recorded earthquake in the world to date. If the recalculations by Stein and Okal (2005) are correct, the December 2004 earthquake off the coast of Sumatra, which triggered the massive tsunami in the Indian Ocean, was only slightly smaller, at 9.3. Of course, the geologic settings of Sumatra and Chile are very different from our region of interest, since they both lie on subduction zones, the types of plate boundaries that produce the very largest earthquakes. The North Anatolian Fault is a transform boundary, one that almost certainly cannot produce huge events like these. However, even the largest strike-slip earthquakes can cause devastation over much larger areas than most archaeologists realize. At the time that Blegen, Schaeffer, and their colleagues were having their pivotal argument, the theory of plate tectonics had not yet been introduced, and the causes of earthquakes were poorly understood. That, combined with the short record available for comparison, made such large events seem unrealistic.

The failure of archaeologists to appreciate the widespread damage that a single earthquake can cause is compounded by the tendency in early archaeological literature to misuse the term "epicenter" to mean "zone of destruction." Thus, damage from a single earthquake would sometimes be erroneously described as having many "epicenters," when, in reality, this usually just meant that there were many places where construction density was high enough to suffer severe earthquake damage. The terminology was confusing.

In any case, based on the archaeological evidence, it is unlikely that the damage could have been caused by only one earthquake. The destruction at the end of the Bronze Age was apparently spread over a period of approximately fifty years—say, from about 1225 BC to 1175 BC—so blaming the destruction on earthquakes would require several events in rapid succession. Archaeologists, faced with that requirement, began to feel that the earthquake hypothesis was on shaky ground, so to speak. The feeling that too much was being trusted to coincidence prompted most prominent scholars in the field to abandon earthquakes in favor of more arbitrary human motives.

In the past half-century, however, scientists have learned much about the timing of large earthquakes. As the period for which we have been able to measure earthquake epicenters and magnitudes gets longer—it now stretches for more than a century—we are beginning to see larger patterns emerge, at least in some regions of the world. On March 28, 2005, for instance, only three months after the enormous Sumatra earthquake of December 26, 2004, a very large earthquake of estimated magnitude 8.7 (as of this writing) occurred southeast of the first earthquake's rupture zone. Its focus was on the same plate boundary as the 2004 quake but on a section that did not slip in the first event. Ironically, some of the first detailed scientific analyses of the 2004 Sumatran quake were just appearing in journals when the second earthquake occurred, and one article warned of the continued danger on this southern section of the boundary (Stein and Okal 2005). This is because we now understand that some earthquakes can actually increase the stress on sections of the same fault, or nearby faults, that did not slip initially (King, Stein, and Lin 1994). Thus, one earthquake can increase the chances of another earthquake nearby.

As described in chapter 2, stress on a plate boundary accumulates gradually over a prolonged period, sometimes as long as several hundred years. During this time, the fault can be relatively inactive. Sometimes the stress along the entire fault is then released in a single, very large earthquake with numerous aftershocks. However, the strain can also be released in a series of large earthquakes, each one triggering another on an adjacent section of the fault a few months or years apart. In this way, the fault "unzips" in steps until the stress has been released along its entire length. The process of gradual accumulation and subsequent release of stress would then begin again, leading eventually to another series of earthquakes after centuries of inactivity.

Geophysicists have not agreed on what to call this phenomenon but describe modern examples either as "earthquake sequences" (e.g., Ambraseys 1970), "earthquake migrations" (Mogi 1968; Roth 1988), "progressive failures" (Stein, Barka, and Dietrich 1997), or, in places where many intersecting faults are involved, "earthquake storms" (Nur and Cline 2000).

An example of a modern "earthquake sequence" occurred on the North Anatolian Fault during the twentieth century (e.g., Ambraseys 1970; Allen 1975; Wood 1996, 225, 226; Stein, Barka, and Dietrich 1997). This thirty-year sequence, which has been well studied by geophysicists, consisted of seven earthquakes—all with magnitudes greater than 5.6—in 1939, 1942, 1943, 1944, 1951, 1957, and 1967. The quakes progressed westward along the 1,000 km-long fault zone, releasing stress that had accumulated over the previous two hundred or more years (Figure 8.9); this sequence might now be extended to sixty-one years in light of the August and November 1999 quakes that continued the earlier rupture pattern.

The amount of displacement, or slip, that occurred on the fault during each of those earthquakes is shown in Figure 8.9. Over a

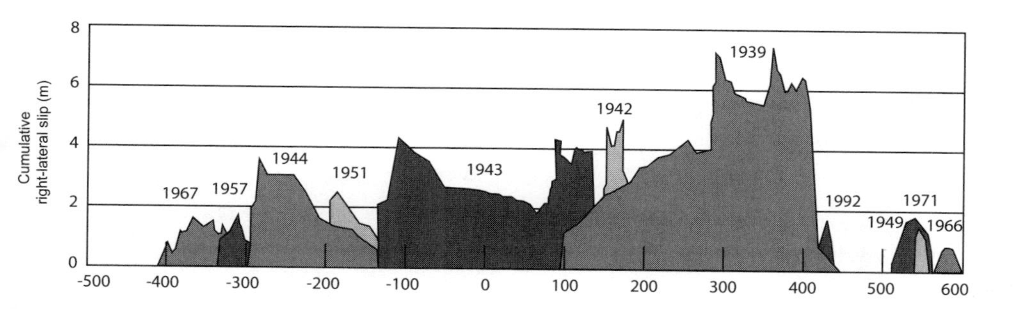

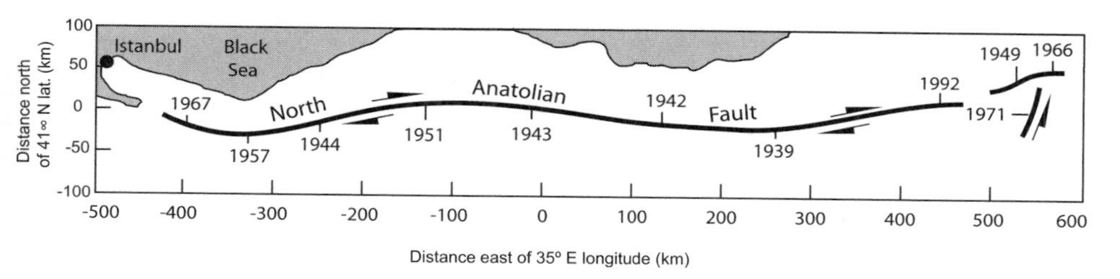

Figure 8.9 The North Anatolian earthquake sequence. The top plot shows the portion of the North Anatolian Fault that slipped with each earthquake, with the height indicating the distance one side moved relative to the other at each point along the fault. The bottom plot shows the location of each earthquake's epicenter (after Stein et al. 1997).

thirty-year-period, the entire plate boundary slipped by 2–4 meters. We know from measured motions of the tectonic plates that the North Anatolian Fault must accommodate a slippage of around 1–2 cm per year, or from 1–2 meters per one hundred years. If the entire fault remains locked for several centuries at a time, a small slip somewhere on the fault will cause a jump in the stress on the rest of the fault. It is as if a wide rubber band in tension were suddenly snipped partway across; the rest of the band would be more likely to break as a result. This is what has happened along the North Anatolian Fault in the twentieth century, and it appears also to have happened in the past.

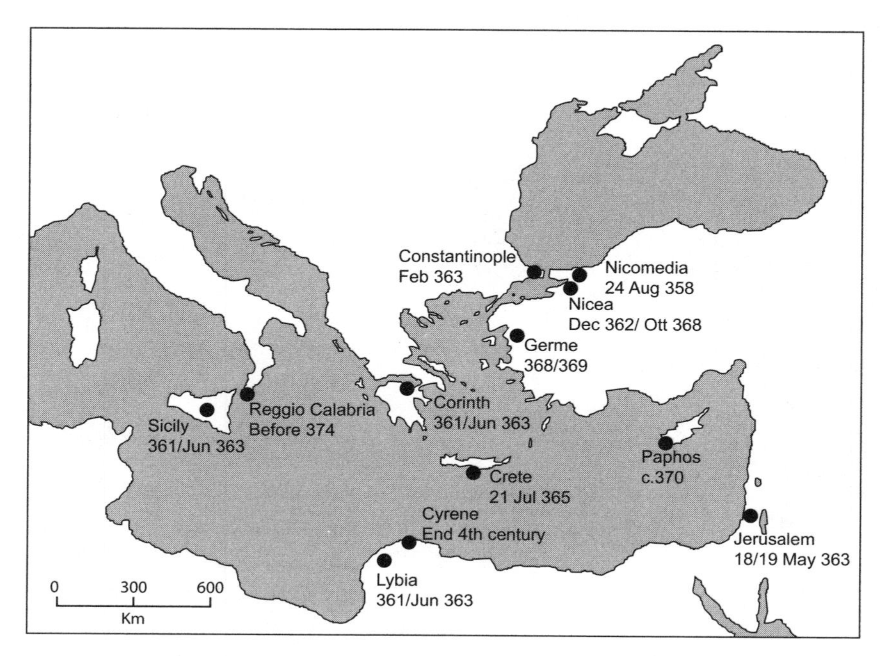

Figure 8.10 The location of the sites damaged in the fourth century AD earthquake "storm" in the Eastern Mediterranean. The dates of the earthquakes are from historical accounts (after Guidoboni et al. 1994.)

A longer earthquake sequence, of more than eighty years' duration, occurred along the North Anatolian Fault from AD 967 to 1050, during which more than "twenty earthquakes of damaging to destructive magnitude occurred" (Ambraseys 1970).

A shorter episode was seen in Greece, where six earthquakes occurred within an eight-year period in a region crisscrossed by interrelated faults. Earthquakes occurred in the areas of Thessaloniki in 1978; Volos in 1980; Corinth, Athens, and Thebes in 1981; and Kalamata in 1986 (Sampson 1996; see also Galanopoulos 1963 on an earlier series from 1953 to 1958). An earlier earthquake storm may also have caused destruction in the mid-fourth century AD (Figure 8.10), when a series of significant earthquakes between AD 350 and 380 resulted in considerable damage at sites in Israel, Cyprus, northwestern Turkey, Crete, Corinth, Reggio Calabria, Sicily, and northern Libya (Guidoboni et al. 1994, 504). This active period was preceded and followed by relatively quiet periods of about three hundred years.

This chapter began by describing how most of the sites destroyed at the end of the Bronze Age have been shaken in modern times by major earthquakes. Now we have also seen that these same areas are subject to earthquake storms or sequences that can wreak destruction over large areas in only a few decades. In view of these observations, it is perfectly plausible that a similar sequence or storm of large earthquakes could have occurred in the Aegean and the Eastern Mediterranean near the end of the Bronze Age, ca. 1225–1175 BC. Thus, Schaeffer's explanation that earthquakes caused the end of the Bronze Age need not depend on a single, unreasonably large earthquake to destroy all the sites he listed. Such widespread earthquake destruction in the region in question can be explained by a mechanism for sequential quakes that is known to act there today.

SKELETAL EVIDENCE

Other evidence supporting the theory that earthquakes were involved in the collapse of the Late Bronze Age comes from skeletons found crushed beneath collapsed buildings in that period, evidence Drews inexplicably dismisses as nonexistent. Reports from site after site vividly illustrate abundant evidence of such remains, with crushed skeletons found in Knossos, Troy, Mycenae, Tiryns, Thebes, Pylos, Gla, Midea, Kynos, Jericho, Ugarit, and Megiddo, all dated (with more or less controversy) to the end of the Bronze Age (Nur and Cline 2000). However, in nearly every case, some scholars reject the notion that earthquakes buried these individuals. In some instances, they argue that in large earthquakes loss of life should have been more widespread. However, except when earthquakes strike at night, one may reasonably assume that most inhabitants manage to escape the collapse of buildings. Further, earthquake survivors generally go to extraordinary measures to sift through the rubble in search of dead relatives or buried valuables. Possibly, then, most of those killed beneath the rubble were extracted and given decent burials, whereas those who could not be recovered remained beneath the rubble.

The evidence is certainly convincing that earthquakes destroyed some or all of the major sites at the end of the Bronze Age; a more complicated issue, however, is whether earthquakes could have destroyed entire societies. Drews obviously does not think so. However, he uses this issue to argue against earthquakes at the end of the Bronze Age, as if the argument that earthquakes were not solely responsible for the societal collapse requires that they did not happen at all.

Nothing in Drews's arguments precludes earthquakes from playing some part, in addition to other factors, in the total collapse. Two scenarios for the prehistoric effects of earthquakes are likely, and these have been repeated throughout recorded history. The first involves a society under military, political, or environmental stress, as was true of the Jews during the Third Temple movement in the fourth century AD. Other examples, which I discuss further in the next chapter, include Spartan society in the fifth century BC, the Portuguese religious establishment in the eighteenth century AD, and Venezuelan revolutionaries in the nineteenth century. In each case, an earthquake or series of earthquakes further destabilized an already ailing society, arguably hastening its collapse. The impact of an earthquake in such times of stress can have very complex results that forever change societies.

The second scenario involves apparently more robust societies, such as the one governed by Herod the Great in the first century BC or the inhabitants of Jericho in the time of Joshua. In Herod's case, an earthquake directly precipitated an attack by the Arabs in what is now Jordan, as the attackers expected that Herod's armies would be vulnerable and demoralized in the earthquake's aftermath. The biblical story of Jericho may be a reflection and reworking of exactly this scenario but with the roles of aggressor and attacked reversed. The aggressors ultimately failed in Herod's case, but they were the victors in the case of Joshua. The point is that the earthquake occurred first, making the attack seem more advisable.

In the absence of written histories, it is hard to tell whether the conflict or the earthquake came first. Indeed, the defensive walls surrounding most cities in antiquity indicate that these ancient societies were always in conflict—the inhabitants fighting either among themselves or with neighbors or foreign raiders—and the occurrence of large earthquakes would leave them exposed and vulnerable to attack.

In the next chapter, I consider instances where earthquakes have made a lasting impression on societies. In each case, earthquakes struck at politically sensitive times, but each involved only one earthquake, in isolation. Because in historical times we have not experienced an earthquake sequence or storm during a period of serious political instability, it is difficult to predict what the result would be.

CHAPTER 9

Rumblings and Revolutions

POLITICAL EFFECTS OF EARTHQUAKES

"How goes it, Bolívar? It seems that Nature has put itself on the side of the Spaniards."

—Monk Don José Domingo Diaz, to Simón Bolívar, March 26, 1812, on the Venezuelan earthquake

"If Nature is against us, we will fight it and make it obey us."

—Simón Bolívar, in response to Diaz

The scene in the Minoan temple described in chapter 5 may illustrate one of societies' most common reactions to devastating earthquakes: a sudden burst of intense religious activity. Although human sacrifice was never depicted in Minoan art, we cannot know with certainty whether this type of religious observance was uncommon in Minoan culture. Perhaps the sacrificial bulls in their paintings and carvings symbolize sacrifice in general or represent the qualities they most admired in healthy males. It seems far more likely, however, that the tableau uncovered in the Arkhanes temple represents a deviation from normal behavior, a religious excess in reaction to an extraordinary catastrophe.

Even in historic times, earthquakes often elicited mass hysteria. On Sunday, February 8, 1750, a modest earthquake shook the city of London, England. The earthquake's felt region was quite small, indicating that the quake was small, shallow, and centered directly beneath the city. Despite causing minimal damage, the event undermined the city's confidence significantly. After all, the most basic daily human activities depend on the assumption that the ground will remain stationary beneath one's feet, and any evidence to the contrary can be quite disconcerting, even today. Especially disturbing to Londoners was the sense that only their city was targeted, with other regions unaffected by the quake. All the same, the disquiet that followed the small earthquake might have passed unmarked by history had a second earthquake not followed.

Exactly four weeks after the first earthquake, on March 8, another one struck London and was so similar to the first that it seemed its twin. The second event was only slightly more damaging than the first, but its extraordinary timing triggered a panic among residents with any religious inclinations. Someone began circulating a prophesy that London would be completely destroyed by a third quake, four weeks from the second, between sunset on April 4 and sunrise on April 5. Despite appeals for calm by many local clergy and administrators, citizens panicked as the prophesied doomsday drew near, and many Londoners with the means to do so—estimates range as high as one-third of the population—packed their belongings and evacuated to the countryside.

When the quake failed to occur, some of those who fled stayed away until April 8, in case the prophet had been mistaken, and the key time interval was a full month instead of four weeks. A general feeling of sheepishness dominated London society after this fiasco, a feeling exacerbated by satirical cartoons and reproving speeches.

Some earthquakes are quickly forgotten, especially when residents wish to forget their own foolishness. Even after devastating earthquakes, people usually rebuild, and the earthquake, or at least the terror of it, fades from memory. It is partly this phenomenon that leads to the formation of tells, and also leads archaeologists to say that even the most destructive earthquakes have little lasting impact on society.

In a politically and economically stable society, the effects may very well be fleeting. In cases where there is an incipient power struggle, however, or even an outright war, the intercession of a sudden earthquake can have far-reaching or lasting effects. Numerous stories in the Bible, for instance, describe God assisting one army or another by making the earth shake. Given the nearly constant history of both conflict and seismic activity in the Holy Land, it is to be expected that earthquakes occasionally coincided with battles. Whether the disadvantaged participants in such battles believed the earthquake was of divine or natural origin, the earthquake itself would be demoralizing. It is certainly conceivable that it could affect the outcome of the battle.

Throughout the Bible, earthquakes are seen as portents, reinforcing one view or another, punctuating victories or punishing the wicked. Whether these events actually changed the course of history can always be argued, but their inclusion in the Holy Book of three major world religions certainly attests to their having shaped society to some degree.

Only five years after those embarrassing little London earthquakes, an earthquake in Lisbon, in 1755, also shaped society and religion. The Lisbon earthquake led to an eventually brutal confrontation, not on the field of battle but between the church and the secular government of the city. It was a rehash of one of society's oldest power struggles—that between kings and priests—and, as it turned out, that battle was to have profound repercussions for Western society.

LISBON, 1755

The November 1, 1755, Lisbon earthquake may have had more influence on Western thought than any other quake, for it was "an event which indirectly changed men's thinking about their own place in nature" (Besterman, 1969). Because such earthquake catastrophes are rare in Western Europe, this one stimulated great scientific interest for centuries afterward (Richter 1958), kindling fierce debates between "common sense" and "natural law," and between theologians and Enlightenment philosophers.

Centered somewhere off the coast of Portugal, the earthquake occurred long before the invention of magnitude or intensity scales. However, modern scientists have tried to estimate the magnitude retrospectively. Charles Richter (1958) suggested that the magnitude of the 1755 Lisbon quake could "scarcely have been less than 8½ and may have approached 8¾." According to Jan T. Kozak of the Institute of Rock Mechanics at the Czech Academy of Science and Charles D. James of the National Information Service for Earthquake Engineering at the University of California,

> The oscillation of suspended objects at great distances from the epicenter indicate an enormous area of perceptibility. The observation of seiches as far away as Finland suggest a magnitude approaching 9.0. Precursory phenomena were reported, including turbid waters in Portugal and Spain, falling water level in wells in Spain, and a decrease in water flow in springs and fountains. (Kozak and James 1998)

Arch C. Johnston (1996) speculated that the earthquake's magnitude was between 8.3 and 9.1, the largest known oceanic event not associated with a subduction zone, and faulting may have extended as far as 60 kilometers beneath the seafloor. We will never know the exact magnitude of the quake, but, by all accounts, it was so large that the shaking affected most of Europe and stretched to the North African countries of Morocco and Algiers. The enormity of the area where the earthquake effects were felt caused general alarm and astonishment, and a great wave of speculation, scientific and otherwise.

Lisbon was one of the economic capitals of Europe at the time of the quake, important in trading, art, and finance (Figure 9.1). Militarily powerful, and a seat of enormous religious influence under the Roman Catholic Church and its Inquisition, the city seemed invulnerable. Its people were some of the most affluent and fervently religious in all of Europe at the time. Lisbon was also home to a thriving textile trade and a large number of cultural treasures, including numerous collections of art and a vast number of churches and cathedrals.

The earthquake struck at 9:30 AM on November 1, the day that marks the observance of All Saint's Day in the Roman Catholic

Figure 9.1 Lisbon before the 1755 quake. The buildings of the Royal Palace of the Corte Real family and Terreiro do Paco situated near the Tagus River were later destroyed by the earthquake and fire (Kendrick 1955).

Church. Many of Lisbon's approximately 275,000 residents (Kendrick 1956, 34) were attending church services when the earthquake struck. More than twenty parish churches and cathedrals collapsed, causing many deaths; the combination of high ceilings and unreinforced masonry arches and vaults was deadly. An estimated 50,000 to 60,000 townspeople died (Maxwell 1995; Kendrick 1956; Walker 1982; Richter 1958). The figures are not precise, as Earnest Zebrowski Jr. (1997) points out: "Given the fire and the seismic sea waves, body counts were not practical, nor was such a gruesome accounting procedure high on anyone's list of priorities." The earthquake captured the public imagination for a long time afterward, and the destruction of Lisbon was the subject of countless paintings and drawings. One artist's rendering of some of the damage to the city is shown in Figure 9.2.

Many of those who were not trapped in the collapsed houses of worship fled to the harbor on the Tagus River, trying to escape what they saw as a doomed city. Unfortunately, a tsunami had been triggered by movement of the seafloor during the shaking,

Figure 9.2 Lisbon after the earthquake. A 1757 drawing, made by Messrs Paris et Pedegache, of the Praca de Patriarcal (Patriarchal Square) destroyed by the 1755 earthquake and the fire that followed (courtesy Jan Kozak Collection, Earthquake Engineering Research Center, University of California, Berkeley).

and the great wave thundered up the river from the sea. Despite being impeded somewhat by the river bar, it reportedly reached some 15 to 20 feet high, in three surges. More than a hundred people were swept from the banks and drowned, and all the light structures on the harbor were destroyed (Figure 9.3). Many of the ships in the harbor were smashed, sunk, or sucked out to sea by the receding waves.

Lisbon's misery did not end there, however. Hundreds of fires were triggered by the upset of candles, hearths, and lamps in the city, and most of the city burned to the ground over the six days that followed. According to the British Historical Society of Lisbon (1990):

> As soon as it grew dark another scene presented itself, little less shocking than those already described. As soon as it grew dark the whole City

Figure 9.3 The Lisbon harbor after the 1755 earthquake (courtesy Jan Kozak Collection, Earthquake Engineering Research Center, University of California, Berkeley).

> appeared in a blaze, which was so bright that I could easily see to read by it. It may be said without exaggeration, that 'twas on fire at least in an hundred different places at once, and thus continued burning for six days together, without intermission, or the least attempt being made to stop its progress. It went on consuming everything the earthquake spared.

Although the earthquake and tsunami were responsible for most of the deaths that occurred that day, the fires caused much of the economic destruction, consuming the priceless works of art, documents, and architecture that had made the city famous in its day. Damage totaled more than forty-eight million Spanish dollars, equivalent at the time to twelve million pounds Sterling. When the fires burned out—for few were extinguished by firefighters—the city was littered with tens of thousands of human corpses and many dead animals as well.

RELIGIOUS FAITH IN A PREDICAMENT

There were so many people killed in the collapsed churches that many believed the punishment of God was upon them. They appealed to their religious leaders, asking why such horrors would be visited upon the faithful. The response from the religious establishment and the religious fringe was widely varied. Some said that God was appalled by the wickedness of the people of Lisbon and was punishing them for everything from materialism to idolatry. Another response was that God was angry with the world in general, and that the people of Lisbon had been singled out as a warning, not because they were particularly wicked but because God especially loved them and was concerned that they repent for the sake of their souls. None of these explanations found widespread acceptance, however. The Portuguese were disheartened, and they began to ask themselves if God could really be the loving Father they had been raised to worship.

It was a crisis of faith for Lisbon and, as word of the disaster spread, for the rest of the world as well. The magnitude of the disaster was so great that the usual news routes collapsed. It took weeks for the story to reach London, and when the tale was first told, it was widely disbelieved. Even when the facts were finally known, they were told in so many exaggerated variations that even today it is difficult to know the truth. The world's reaction was passionate and mixed; there was great sympathy for the people of Lisbon, but, at the same time, preachers throughout Europe went wild. Some smugly pointed out that the Portuguese, with their icons and images of saints, had ignored the commandment against graven images. The earthquake, they said, was an admonition from God telling Christians that they needed to return to Mosaic Law. Others warned that the Lisbon quake was a portent of global disaster to follow, that the Portuguese were only minor sinners and the real destruction would strike closer to home. The religious fervor surrounding the whole affair was unparalleled in European history. In fact, world history leading up to the earthquake, along with the events following the disaster, combined to make this not only a religious crisis but also one of the major turning points in popular thought.

One of the principals in this drama was Sebastião José de Carvalho e Mello. In 1770, he was made the Marquês de Pombal and history generally remembers him by the name Pombal. At the time of the Lisbon earthquake, he was Minister of War and Foreign Affairs under King José I. After the earthquake, the king effectively made Pombal dictator, and it was because of the latter's practical and immediate command of the situation that most of Lisbon's misery was confined to the physical effects of the earthquake and the subsequent fires.

When the earthquake and the conflagration had subsided, Pombal's military sense of organization spurred him to action. His first concerns were disposing of the dead and feeding the living. He arranged camps for survivors, providing them with safe food and water. Meanwhile he turned to the problem of the tens of thousands of corpses in the city.

Although the doctors of the day did not understand the mechanisms of disease transmission, there was some concern that the presence of many corpses and animal carcasses rotting in the streets could not be healthy for the survivors. The overriding fear after the earthquake was that a host of illnesses would sweep through the already decimated populace. To prevent this, Pombal recommended to the Patriarch of the Church, Cardinal José Manuel, that the dead bodies from all over the city be gathered, weighted, loaded on a barge, and sunk on the ocean side of the bar in the Tagus River. The cardinal agreed, and presumably the deed was done, along with heroic burial efforts on other fronts. This mass disposal was perhaps the most expeditious way to deal with the public sanitation problem in the city, but although the Patriarch sanctioned it, the process nonetheless was carried out in secrecy, for it would certainly have angered many of the lesser clergy in the city.

Having met the immediate needs of the city, Pombal began to make plans for reconstruction, but many dissenting voices among the clergy counseled prayer and repentance, not community action. At least two officially sanctioned penitential processions through the city were mounted, one of them featuring the royal family itself. Many clerics in the city preached an extreme and self-destructive brand of repentance, which, Pombal feared, would provoke a religious frenzy and interfere with what he saw as the more immediate problem of persuading people to perform their civic duty and minister to their neighbors.

Pombal's conflict with the church did not begin with the earthquake. He already was at odds with the Jesuits, who he thought had entirely too much political power in Portugal and its colonies. Perhaps partly because of this preexisting conflict, he saw the Jesuits as the ringleaders of the counterproductive religious fear-mongering.

A particular thorn in Pombal's side was the Italian-born Jesuit missionary Gabriel Malagrida. Malagrida had great personal influence with the Portuguese royal family, and his missionary exploits in the Portuguese colony of Brazil had made him a popular figure with the citizens of Lisbon. Whereas Pombal wanted to emphasize the natural causes of earthquakes, Malagrida published a pamphlet detailing his own contrary view: "Learn, O Lisbon, that the destroyers of our houses, palaces, churches, and convents, the cause of the death of so many people and of the flames that devoured such vast treasures, are your abominable sins" (quoted in Kendrick 1956).

The pamphlet, probably a printed version of the sermon Malagrida had been preaching ever since the quake occurred, held that the catastrophe was caused by the wrath of God, and that it was "necessary to devote all our strength and purpose to the task of repentance." This was exactly opposed to Pombal's wish to focus the people's main energy on repair and the maintenance of order. Undoubtedly, Malagrida's pamphlet was a direct challenge to Pombal and infuriated him.

Malagrida's zeal in spreading his teaching eventually made him a public menace in the eyes of the government. However, the Roman Catholic Inquisition to uncover and punish heresy was in full force at that time, and there was widespread fear of directly opposing the Church. The political machinations were complicated and took years to resolve, but, in the end, the Jesuits were expelled from Portugal and the state triumphed over the Church. Malagrida himself did not escape lightly. The government remanded him to the very Inquisition that had previously protected him, and he was sentenced as a heretic based on blatantly false charges and publicly executed in 1761. Pombal had won, and he rebuilt the city.

This clear-cut control over the highest authority of the Church by an arguably secular government had never occurred before. Portugal was a strictly Catholic country, yet Pombal's actions were driven entirely by secular expediency. Some historians point to this as a major turning point that ushered in the modern political era in the West. As Judith Shklar (1990) pointed out:

> What makes it such a memorable disaster is not the destruction of a wealthy and splendid city, nor the death of some ten to fifteen thousand people who perished in its ruins, but the intellectual response it evoked throughout Europe. . . . From that day onward, the responsibility for our suffering rested entirely with us and on an uncaring natural environment, where it has remained.

A BLOW TO OPTIMISM

Evidence of the profound impact this earthquake had on European thought can also be found in the literature and philosophy of the day. The cruelty of the Lisbon disaster was a serious blow to the philosophy of Optimism, one of the most popular beliefs of the day, articulated and defended by the early eighteenth-century thinkers Gottfried Wilhelm von Leibniz (1646–1716) and Alexander Pope (1688–1744). Optimism, the idea that everything in the world worked at its deepest level toward the general good, was embodied in Pope's phrase, "Whatever is, is right," from his *Essay on Man*, and Leibniz's notion that this world is "the best of all possible worlds." This was, of course, a matter of some debate even before the earthquake, but the cruelty of the Lisbon disaster was a serious blow to the idea of Optimism. A letter from

François-Marie Arouet de Voltaire to the banker M. Tronchin, dated November 24, 1755, two weeks after the earthquake, illustrated the cynical attitude brought on by the tragedy:

> My dear sir, nature is cruel. One would find it hard how the laws of movement cause such frightful disasters in the best of possible worlds. A hundred thousand ants, our fellows, crushed all at once in our ant-hill, and half of them perishing, no doubt in unspeakable agony, beneath the wreckage from which they cannot be drawn. Families ruined all over Europe, the fortunes of a hundred businessmen, your compatriots, swallowed up in the ruins of Lisbon. What a wretched gamble is the game of human life! What will the preachers say, especially if the palace of the Inquisition is still standing? I flatter myself that at least the reverend fathers inquisitors have been crushed like the others. That ought to teach men not to persecute each other, for while a few holy scoundrels burn a few fanatics, the earth swallows up one and all. (quoted in Besterman 1969)

The Lisbon earthquake compelled Voltaire to mock the arguments of the Optimists. He published a massively pessimistic poem on the Lisbon earthquake titled "Poeme sur le desastre de Lisbonne" ("Poem on the Lisbon Disaster"), disputing the image of a benign and all-loving God who created "the best of all possible worlds":

> Why suffer we, then, under one so just?
> There is the knot your thinkers should undo.
> Think ye to cure our ills denying them?
> All peoples, trembling at the hand of God,
> Have sought the source of evil in the world . . .
> But I, I live and feel, my wounded heart
> Appeals for aid to him who fashioned it. (Voltaire 1911 [1756])

If the earthquake that had caused such pain and devastation could have been averted, but was not, then, according to Voltaire, the world could never be restful again, for evil must exist. He continued his reflection on the dreadful consequences of the earthquake and the idea that Lisbon had been randomly destroyed:

> Did fallen Lisbon deeper drink of vice
> Than London, Paris, or sunlit Madrid?
> In these men dance; at Lisbon yawns the abyss

According to the Bible, the citizens of Nineveh were given an ultimatum: repent within forty days or be destroyed. Why did a benevolent God not extend the same mercy to the people of Lisbon? Were there no harmless forewarnings to send before this disastrous quake? Why had He chosen such a religious day, so that the bulk of those killed died at worship?

Voltaire's poem, and another poem he wrote shortly afterward, greatly disturbed Jean-Jacques Rousseau, another prominent figure of the Enlightenment. Rousseau wrote a letter in response, dated August 18, 1756, in which he complained,

> Your last two poems have reached me in my solitude. . . . All my complaints are . . . against your poem on the Lisbon disaster, because I expected from it more worthy evidence of the humanity that apparently inspired you to write it . . . but you so burden the list of our miseries that you further disparage our condition. (Leigh 1967)

Rousseau argued that man's insistence on congregating in unnatural cities, instead of leading the sort of dispersed, pastoral life that he felt was "natural," was the cause of the tragedy at Lisbon. He faulted Voltaire for charging divinity with the miseries listed in the poem, and pointed out that it was man's responsibility:

> Without leaving your Lisbon subject, concede, for example, that it was hardly nature who assembled there twenty-thousand houses of six or seven stories. If the residents of this large city had been more evenly dispersed and less densely housed, the losses would have been fewer or perhaps none at all. Everyone would have fled at the first shock, and would have been seen two days later, twenty leagues away and as happy as if nothing had happened. . . . How many unfortunates perished in this disaster for wanting to take—one his clothing, another his papers, a third his money? (Rousseau, quoted in Besterman 1969)

Rousseau contended that God had chosen to order the natural world in the best possible way, but man's actions were at cross-purposes with God's. Effectively he supported the "all is for the best" philosophy, at least where natural phenomena were concerned.

Rousseau's argument, along with continued public support of Optimism, prompted Voltaire to expand his harsh opinion in prose a few years later. His black satire, *Candide,* was considered by many historians to be the final nail in the coffin of Optimism. Voltaire repeatedly caricatured Leibnitz's "best of all possible worlds" philosophy in the character of Candide's optimistic tutor, Dr. Pangloss. Here is what Pangloss has to say about the 1755 Lisbon quake:

> This earthquake is no new thing, the city of Lima in America experienced similar shocks last year; the same causes, the same effects; . . . For nothing is better, for if there is a volcano under Lisbon, it could not be elsewhere, for it is impossible that things should not be where they are, for all is well. (Voltaire 1759; English ed. 1998, 120–121)

Like many ancient earthquakes, the Lisbon event caused great physical devastation and human suffering. However, it was unusual in that it also changed the course of world events. By striking at a time when there was a particularly delicate balance of power between church and state, and between science and religion, the earthquake tipped the scales and changed society around the world. The old power would never again enjoy such dominance over the new. There has been no other such event in recent times, when natural and political factors combined to such a far-reaching effect. However, several earthquakes have had smaller-scale political implications, and have influenced the local power balance in regions affected by the quakes.

SPARTA, 469–464 BC

Some of the earliest, reliable written evidence for earthquakes coinciding with war is from the fifth century BC, in Greece. Thucydides (1910) described how the Spartans had promised to help the Thasians invade Attica (without the knowledge of Athens) when a terrible earthquake devastated Sparta and the surrounding countryside, preventing them from keeping their promise. Sparta was soon busy trying to quell an uprising by the Messenians and the Helots of the region, who saw that Sparta was already half-destroyed and chose that opportunity to revolt.

Guidoboni, Comastri, and Traina (1994), of the *Istituto Nazionale de Geofisica* in Italy, offer an excellent overview of historical references to this earthquake, as well as some discussion of conflicting modern interpretations. These authors cite Diodorus as giving the following detailed account of the damage from the earthquake:

> During the year, a great and extraordinary catastrophe struck the Spartans, for violent earthquakes occurred in their city, so that houses completely collapsed and more than twenty thousand Spartans were killed. Since the city was shaken and many houses collapsed over a long period of time without respite, many people were caught and crushed by collapsing walls, and a great deal of household property was destroyed. They suffered this disaster because, it seemed, some god was venting his anger on them, but they also had to face dangers of human origin, as is set out below. Although the Helots and Messenians were enemies of the Spartans, they had remained quiet up to now, because they were afraid of Sparta's overweening power; but when they saw that a majority of Spartans had perished in the earthquake, they felt contempt for the few survivors. So they came to an agreement and waged a joint war against the Spartans.

Geophysical studies in Greece independently support the occurrence of this earthquake; field studies and satellite imagery have identified a fault scarp 10–12 meters high and extending for about 20 kilometers, passing within only a few kilometers of the site of ancient Sparta (Armijo, Lyon-Caen, and Papanastassiou 1991; Benedetti et al. 2002; Higgins 1996).

This earthquake was not the only one to strike during the Peloponnesian War, nor the only natural disaster. As described in chapter 4, Thucydides tells of several other earthquakes occurring during the sixth year of the war:

> The next summer, the Peloponnesians and their allies set out to invade Attica under the command of Agis, son of Archidamus, and went as far as the Isthmus, but numerous earthquakes occurring, turned back

> again without the invasion taking place. About the same time that these earthquakes were so common, the sea at Orobiae, in Euboea, retiring from the then line of coast, returned in a huge wave and invaded a great part of the town, and retreated, leaving some of it still under water; so that what was once land is now sea.

Thucydides goes on to describe other coastal towns and islands affected by this inundation, as well as an earthquake that simultaneously threw down the town hall and other buildings in Peparethus. He writes,

> The cause, in my opinion, of this phenomenon must be sought in the earthquake. At the point where its shock has been the most violent, the sea is driven back, and suddenly recoiling with redoubled force, causes the inundation. Without an earthquake I do not see how such an accident could happen.

Perhaps Thucydides' most interesting observation comes from his first introduction of the earthquake: "Old stories of occurrences handed down by tradition, but scantily confirmed by experience, suddenly ceased to be incredible; there were earthquakes of unparalleled extent and violence."

Thus, we have a rare glimpse from ancient times into the even more ancient past, when events were recorded not by written history but by oral traditions that would have been lost to history were it not for this brief mention by Thucydides. Similar events, of course, had occurred in the past, but the people of Thucydides' time had not believed tales of them until they recurred in their own day. The effects of the earthquakes were complex and far-reaching, and probably were partly responsible for the unparalleled length of the Peloponnesian War. How might things have turned out differently had the Spartans not been crippled by the earthquake near the outset of the war?

VENEZUELA, 1812

In more modern times, military undertakings also have been affected by earthquakes. On March 26, 1812, for instance, a large earthquake struck Venezuela. It happened on Holy Thursday, the day before Good Friday of Easter week, when most Venezuelans went to church to participate in the traditional processions celebrating that day. The day also marked a milestone in Venezuela's struggle for independence from Spain; exactly two years earlier the city council of Caracas had formally deposed the Spanish viceroy and appealed to Britain for aid. Also at the time of the earthquake, the royalist general Juan Domingo de Monteverde was actively campaigning to regain Spanish control of the country.

Scientists disagree whether on that day a single massive earthquake or up to four smaller events occurred, but there was heavy damage from Caracas to Merida, a distance of about 500 km. The city of Caracas, a stronghold of the revolutionaries (who called themselves the "Patriots"), was nearly destroyed. Heavy damage was also reported in the Patriot-controlled cities of Barquisimeto, La Guaira, Merida, San Carlos, and San Felipe. Fire followed the quake, engulfing many structures and the people trapped inside. Various sources give the total number of dead in the affected towns and villages as ranging from 20,000 to 120,000 (Ganse and Nelson 1981). Caracas alone lost about 10,000 people.

Damage was relatively light in Coro, Guayana, Maracaibo, Puerto Cabello, and Valencia, all towns loyal to Spain. Predictably, this seeming inequity in the destruction (related, of course, to the geography of the fault system in Venezuela) provoked superstition and fear in the general populace. The survivors feared that it signified God's displeasure with the independence movement, a view supported by the Church, which had remained sympathetic to the Spanish throughout the conflict. The earthquake shook the confidence of the Patriot forces, leaving them vulnerable and disheartened, and rumors abounded. According to Robert Harvey (2000), it was said that in Barquisimeto alone a deep crevice swallowed up about 1,500 Patriots.

In Caracas, Simón Bolívar, Latin America's "*El Liberator*," was awakened from his afternoon nap by the deafening sound of crumbling structures. Leaving his wrecked residence, he hurried with associates to help the survivors. One overzealous monk, Don

José Domingo Diaz, preaching at the central square of San Jacinto, caught Bolívar's attention. In a heated exchange, the monk, a supporter of the Spanish king, ridiculed Bolívar: "How goes it, Bolívar? It seems that Nature has put itself on the side of the Spaniards." Bolívar responded angrily, "If Nature is against us, we will fight it and make it obey us" (Harvey 2000).

As the city stood in ruins, the Church declared the catastrophe an act of God against the nationalist uprising. The royalist clergy, seeking to instill fear in the people of Venezuela, preached that the quake was heavenly revenge against the adversaries of the Spanish king, equating the destruction to the biblical punishment of Sodom and Gomorrah. They called for contrition to regain God's favor, and they condemned the Patriots for not honoring the anointed Spanish king:

> God has wished to punish the patriots and the dissolute morals of the Venezuelans. The anniversary of [the Royalist governor] d'Emperán's banishment had been chosen by God for the execution of His justice. Two years, and the chastisement has proved terrible. (Vaucaire 1929)

The earthquake paved the way for Spanish forces to invade most of the quake-destroyed towns with little resistance. However, a major battle for Valencia erupted between Monteverde and the Patriots under Bolívar and Francisco de Miranda. On March 23, the newly appointed commander-in-chief Miranda was given dictatorial authority in the hope that he could restore order and boost morale in the republic. Instead, there was greater chaos and demoralization, and the Patriots suffered an enormous setback in their struggle for independence. Moreover, Miranda miscalculated his defensive tactics toward Monteverde and could not effectively organize his distressed army. He surrendered on July 30, 1812, and the Spanish reestablished control of the country.

Upon the failure of his First Republic, Bolívar fled into exile, a move that had huge repercussions for South American politics. In exile, Bolívar worked tirelessly against Spain, extending his liberation efforts to other Latin American countries. Had he succeeded in his first attempt to free Venezuela, Bolívar might not have also led Columbia, Ecuador, Peru and Bolivia to freedom from Spanish rule.

Bolívar saw clearly at the time that the earthquake had been a major impediment to his success in Venezuela. In his *Cartagena Manifesto* of December 1812, aimed at scrutinizing the missteps of the republic, Bolívar described the earthquake as one immediate reason for the republic's collapse:

> The earthquake of 26 March was, without a doubt, both physically and mentally destructive, and it can rightly be called the immediate cause of Venezuela's downfall, but this occurrence would not have had such fatal effects if, at the time, Caracas had been governed by a single authority, which, acting quickly and energetically, could have repaired the damage without hindrance or rivalry, whereas, in fact, interference with the measures to be taken compounded the disaster to such an extent that it became irremediable.
>
> If Caracas, instead of a weak and insubstantial federation, had established a straightforward government, as the political and military juncture required, you would exist today, Venezuela, and enjoy your freedom!
>
> After the earthquake the Church played a very considerable part in the rebellion of small towns and villages, and in allowing our enemies to enter the country, sacrilegiously abusing the sanctity of its ministry to the advantage of the ringleaders of civil war. (Bolívar 1983 [1812])

Again, religious fervor and political instability surrounding an earthquake led to these factors having a wider influence than they might otherwise have had, not only for the earthquake but also for the revolution.

TOKYO, 1923

Japan, lying at the conjunction of the Pacific plate, the Eurasian plate, and the Philippine micro-plate, suffers from frequent large earthquakes. Japanese society has learned to deal with this seismic threat, and today it has one of the world's most stringent building codes for quake-resistant structures. This was not true, however, when the great Kanto earthquake struck on September 1, 1923; in fact, Japan's preparedness today is largely the result of that devastating quake. Centered in the Sagami Bay Trough, nearly 70 kilometers from Tokyo, with an estimated magnitude of 7.9, it caused tremendous damage to the city (Figure 9.4), filling the streets with debris from collapsed houses, buried vehicles, and dead and injured people (Figure 9.5).

Within minutes, fires had broken out all over the city, sparked largely by kerosene cook stoves and charcoal braziers that were commonly used at the time (James and Fatemi 2002). The narrow streets choked with debris made it difficult for firefighters to reach the blazes, and the sheer magnitude of the destruction prevented a rapid response to the emergency. In residential areas, the situation was made worse by the extreme flammability of the traditional wood and paper homes. Fires burned continuously for forty-two hours, fanned by high winds, and eventually almost fifteen square miles, or two-thirds of the city, burned (Figure 9.6). In all, three-fourths of the buildings were destroyed by the combined effects of the earthquake and the fires (Seidensticker 1983). Approximately 140,000 people died in the disaster (EQE International 1995).

The loss of life in this earthquake was tremendous (Figure 9.7), with 143,000 people in Tokyo and Yokohama missing and dead, many more injured, and 2 million people left homeless. Beyond its terrible physical consequences, this event also exacerbated existing racial and political tensions in the city. After the earthquake, perhaps partly because residents did not know what to expect after an earthquake, and did not know that fires often followed large quakes, unfounded rumors spread that Koreans and Chinese living in Tokyo had poisoned wells and started fires everywhere. A report surfaced that "the Koreans were up to mischief everywhere, throwing bombs and setting fire to the still remaining parts of the city" (Aurell 1923). Newspaper offices suffered tremendous damage from the fire, which may be why they did not verify the facts before publishing the accusations. The newspaper headlines the next day read, "Koreans Are Throwing Poisons into Wells,"

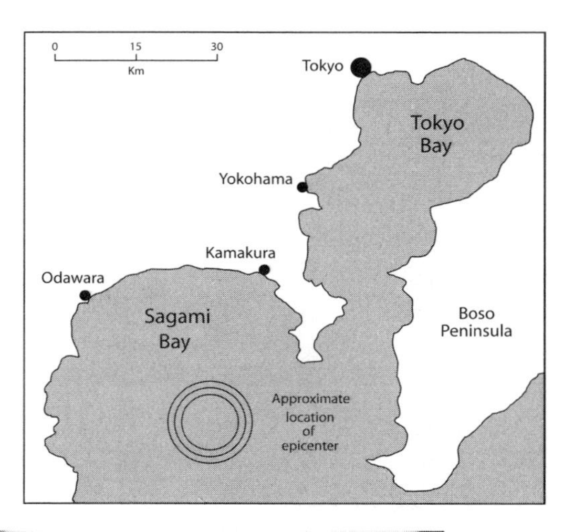

Figure 9.4 A map of the area affected by the 1923 Kanto earthquake, showing severely damaged cities. The epicentral area was located beneath the waters of Sagami Bay (after Bolt 1993).

Figure 9.5 Damage after the 1923 Kanto earthquake in Tokyo, Japan. Pictured here is what remained of the Kawasaki Denki Company building, which was made of reinforced concrete (courtesy Karl V. Steinbrugge Collection, Earthquake Engineering Research Center, University of California, Berkeley).

Figure 9.6 In search of a safe haven, survivors carrying only few of their possessions packed Ueno train station, as smoke from fires throughout the city darkened the sky (courtesy Sykes Kanto Collection, Earthquake Engineering Research Center, University of California, Berkeley).

"Three Thousand Wild Koreans Are Coming All the Way from Kanagawa to Attack Tokyo at Any Moment," and "Many Socialists Are Setting Fires" (Chamoto 1984).

The Tokyo Metropolitan Government reported, in 1995, that the misinformation could have been corrected through a faster mode of communication, but the earthquake had damaged the telegraph and telephone systems. The report suggests that radio broadcast, established only in 1925, might have had a positive effect had it existed in Tokyo at that time.

No efficient way existed to quell suspicions. The police and army, in fact, contributed to the rumor mill (Nakajima 1973) and even participated in the killings that followed (Waley 1991). When order was finally restored, by some estimates more than six thousand Chinese and Koreans had been massacred. The Tokyo Metropolitan Government report of 1995 puts the figure at twenty-five

Figure 9.7 Ashes of about thirty-three thousand victims of the Great Kanto earthquake and fire lie on the grounds of a shrine in Akasaka district (courtesy Sykes Kanto Collection, Earthquake Engineering Research Center, University of California, Berkeley).

hundred, admitting that the number is speculative, since it was impossible to determine whether many died because of the massacre or in the earthquake itself. Moreover, according to the report, "among those killed were Japanese people from the northern part of Japan, Tohoku, and from the southern island of Kyushu, who were mistaken for Koreans because of their accents." Rev B. S. Moore (1923), who witnessed one horrific episode, described the whole event as "a typical feudal war."

In November 1999, a ninety-one-year-old Korean woman, Mun Mu Son, an escapee of both the quake and the slaughter in Tokyo's Shinagawa city, officially reported her recollections of the massacre. Some Japanese, she recalled, burst into the residence armed with swords and fire hooks, shouting, "Wipe out Koreans!" and repeating the rumors about fires and poisoning before killing her father and friends. Eighty years after the fact, Korean survivors and their families called for a complete investigation of the crimes committed in the wake of the great Kanto earthquake.

The same sentiment was echoed in August 30, 2003, when the Japan Federation of Bar Associations recommended to the prime minister that the government acknowledge its responsibility, apologize to survivors and their families, and investigate "the slaughter in its entirety." Although, as Moore (1923) pointed out, the Koreans were innocent of the poisonings and arson for which they had been blamed, one of his sources, K. E. Aurell (1923), suggested that both sides might have spun out of control in the aftermath: "We are inclined to believe there were bad elements of the Japanese behind it. Yet, it does seem true that some of the Koreans took advantage of this occasion to give vent to their feelings of resentment against the Japanese. Time may make that clear."

All this bears testimony to the confusion that typifies the aftermath of a natural disaster. However, much of the chaos and violence might have been avoided with better preparedness. Had it been publicly known that fires commonly occur after earthquakes, the Japanese might have been less quick to blame the Koreans for starting the fires. Had there been an emergency plan in place before the earthquake, there might have been less panic in the aftermath.

Because of the turmoil following the 1923 earthquake, the Japanese government took steps toward educating the populace about earthquake hazards and safety. In 1960, September 1, the anniversary of the Kanto earthquake, was established as Disaster Prevention Day in Tokyo. On this date every year, the city conducts massive earthquake-preparedness exercises. As a result, residents of Tokyo are very aware of the threat of earthquakes there.

Besides the political and racial turmoil it spawned, the great Kanto earthquake also caused considerable economic damage, totaling more than $68 billion, measured in today's currency. However, this disaster did not destroy or even severely cripple Japan's economy at the time. Tokyo was rebuilt "at a faster pace" over the next seven years, and the result was a more modern city, with wider streets and better anti-seismic and fire-resistant construction. It is difficult to imagine the effect this earthquake might have had if it had struck a few years later, when Japan was entering World War II.

The effectiveness of Japan's resulting preparedness has been put to a partial test. On January 17, 1995, a much smaller earthquake struck Japan. With a magnitude of only 6.9, the great Hanshin earthquake (widely known as the Kobe earthquake), was a startling wake-up call to the Japanese. Despite Japan's outstanding seismic building codes and public earthquake awareness, the death toll for this quake was more than five thousand, about thirty-five thousand people were injured, and almost three hundred thousand were left homeless. The fiscal loss was around U.S.$147 billion (EQE International 1995). This devastating earthquake showed that, even with modern anti-seismic measures in place, rescue efforts can still be hampered by crowds of people in the streets and by rubble from buildings in the densely built city. A study conducted by Risk Management Solutions (1995) estimates that, had this earthquake occurred beneath more densely populated Tokyo, the economic consequences would have been astounding. Damage to residential and commercial property would total somewhere between $1 trillion and $1.6 trillion in U.S. dollars. Losses resulting from business interruption would be just slightly more, yielding a total loss of $2.1 to $3.3 trillion. The larger amount is 70 percent of Japan's gross national product. The estimate does not include some losses that are difficult to predict, including damage to utility lines, removal of earthquake rubble, and cleanup of toxic spills (Risk Management Solutions 1995).

Such an earthquake today would be truly devastating to the Japanese people. Only 1 percent of the total amount of property at risk is covered by insurance because of government limitations on earthquake coverage. There is no comparable threat from earthquakes in the United States, simply because it is so much larger, and a much smaller proportion of its people and resources would be affected by even a major earthquake.

Katsuki Takiguchi, a professor at the Tokyo Institute of Technology, stated, just after the Kobe earthquake, that the lesson learned from the event was that the country "was not earthquake-proof," a statement that startled those around the world who thought Japan was far ahead of everyone else in earthquake preparedness (Begley et al. 1995). Tsuneo Katayama (1996), president of the National Research Institute for Earth Science and Disaster Prevention based in Tokyo, wrote:

> For almost 50 years a seismically quiescent period had continued in Japan's urbanized areas, and this period was the golden time for the advancement in earthquake engineering research and technology in Japan. These two independent facts were misused to establish the safety myth, a myth that the time had come when Japanese structures would not collapse even subjected to strong earthquakes. . . . After having successfully constructed many infrastructures including the Tokaido shinkansen bullet train and the Tokyo metropolitan expressways, engineers seemed to have acquired too much confidence in what they had been doing. Engineers tended to forget their humbleness towards the nature [*sic*].

Just as Japan has grown since then, so has the rest of the world, and much of the world economy today is intertwined with Japanese business. Of the world's ten largest banks, eight are headquartered in Tokyo. The Tokyo Stock Exchange handles the third-largest trading volume in the world. A huge proportion of the world's high technology and automotive manufacturing is centered in Japan. U.S. domestic markets, which depend heavily on Japanese investment and materials, would certainly be rocked by the sudden withdrawal of Japanese capital needed for earthquake recovery. The eventual repercussions for the world economy would be difficult to imagine.

So far, there has been no repeat of the great Kanto earthquake. Given our knowledge of plate tectonics, the frequency-magnitude relation, and the severity of damage from historical earthquakes in the same region, we know that another such event is likely, and we can estimate its impact. Such a local catastrophic disaster would be a global event, triggering problems far beyond the shaken region itself. Whether a sufficiently large earthquake in Japan or elsewhere could trigger the collapse of a society in today's world is an open question, and one that can only be answered after the fact.

CHAPTER 10

Earthquakes and Societal Collapse

On the impact on humans of natural catastrophes—
"misfortune or injustice"?
—Judith Shklar, *The Faces of Injustice*

Amnon Ben Tor, the archaeology professor from Hebrew University who, as described in chapter 1, participated with me in the documentary *Killer Quakes of the Bible,* put it bluntly when asked about the mysterious destruction of the ancient towns of Dor and Meggido around 1000 BC: "You cannot prove that earthquakes did it" (Rhys-Davies 1994). Regarding the tendency of the Anatolian Fault toward earthquake sequences, Mark Rose (1999), the managing editor of *Archaeology,* declared:

> Such an east-west earthquake series in the Late Bronze Age could have done a number on sites from the Hittite capital, Bogazkoy, to Troy, but I wonder whether such a series of events over a half-century or more would bring everything in the ancient Near East tumbling down—Egypt and Babylonia might have welcomed disruption in the north, if it occurred, and flourished the more for it. You need to prove that [an earthquake occurred] at that time and, beyond that, show how precisely it would have ended civilization as they knew it, from the immediate effects to ripples through political, economic, and social spheres on local and regional levels. Collapse is too vague a word (7.5 on the vagueness scale).

What Is Collapse?

Mark Rose is right; "collapse" is a vague word but one that is used extensively in historical and archaeological literature. Many scholars have struggled to define it more rigorously. The archaeologist Joseph Tainter has listed some of the indicators of collapse that he considers important:

- a lower degree of social differentiation;
- less economic and occupational specialization;
- less centralized control;
- less behavioral control and regimentation;
- less investment in monumental architecture, artistic and literary achievements, and the like;
- less flow of information between a center and its periphery;
- less sharing, trading, and redistribution of resources;
- less organization of individuals and groups. (Tainter 1988, abridged)

One might argue that these strictly societal symptoms, and the definition of "collapse" they imply, are indicative of gradual collapse, which might also be described as "decline."

I find it notable that the physical collapse of buildings and structures is not even mentioned here, only the shift toward less construction of monumental architecture. Yet, the collapse of buildings and the abandonment of collapsed sites, as uncovered in archaeological digs, is a ubiquitous marker of *catastrophic collapse*. Perhaps "catastrophic collapse" should be defined as a separate phenomenon.

The best-known example of a catastrophic collapse is the end of the Bronze Age in the eastern Mediterranean, where the demise of an overarching political and societal system was accompanied by the physical destruction of countless architectural marvels. Drews (1993) offers an exhaustive compilation of sites and archaeological

continued on page 274

evidence for this physical destruction. He also discusses possible causes for the physical collapse:

> The Catastrophe . . . is not invariably attributed to human agency. Although most scholars do hold that the cities were destroyed by men, a minority has explained the Catastrophe . . . as the result of a terrible "act of God." Specifically, . . . [two] archaeologists have claimed that the quakes that destroyed their sites were also responsible for the fires that burned many other famous sites. In this view, the Catastrophe was an "act of God" of proportions unparalleled in all of history.

That Drews described a natural catastrophe as an "act of God" (in quotes) twice in one paragraph reveals his conviction that human agency is an explanation infinitely preferable to the action of cold nature. That the catastrophe was "unparalleled in all of history" also clearly disturbs him, as it does Tainter (1988): "Why, when complex social systems are designed to handle catastrophes and routinely do, would any society succumb? If any society has ever succumbed to a single-event catastrophe, it must have been a disaster of truly colossal magnitude."

Of course, natural catastrophes can be of colossal magnitude and were just that at times. It is not my intention, however, to argue that earthquakes have destroyed healthy, robust societies in the absence of any other influences. Indeed, as I argue in this book, an earthquake-induced catastrophe in the absence of other factors simply does not exist, nor is there any society that has not been influenced by the action of sudden natural phenomena. Perhaps my greatest goal in writing this book is simply to convince archaeologists who work on sites in the earthquake belts of the world that, when uncovering physical destruction—collapsed walls and buildings, colonnades, crushed skeletons, offset keystones—earthquakes should be considered as one of the foremost possible causes of the devastation. The impact earthquakes had on societies is a completely separate question.

I cannot prove that earthquakes ended the Bronze Age, but evidence in many Bronze Age sites indicates that earthquakes occurred at the appropriate time. The evidence is very strong in some cases, but if the best claim is that "it could have happened," then why have I written a book about it?

The answer is that the widespread bias against recognizing the effects of earthquakes and other natural disasters in archaeology and history reflects a deeper, and disturbing, trend in the general population. Just as archaeologists assume that the human dramas that unfolded in the Bronze Age eastern Mediterranean could somehow rule out the occurrence of an earthquake or series of earthquakes, people who live in earthquake country today tend to ignore earthquake risk. Even those who know intellectually that earthquakes indeed pose a threat seem to think that a catastrophe will wait until they get a chance to do those little earthquake-preparedness tasks that they have been postponing. Perhaps they think that earthquakes are the least of their problems and could not occur while they are so busy with their wars, their droughts, their divorces, or their vacation plans. But earthquakes have no regard for the human condition and do not time themselves either for our convenience or with particular regard for our destruction. We need to lose this self-absorption and face the fact that earthquakes are an ever-present, quantifiable risk that we can plan for, if not anticipate, and that can have far-reaching consequences for modern societies, as well as ancient ones.

Archaeologists and historians argue that earthquakes are insufficient to explain the abandonment of towns or cities in the archaeological record. There is no evidence, they say, that any historical earthquake has resulted in such desertions or that earthquakes can change the direction of society; instead societies respond to earthquakes simply by rebuilding and continuing as before. Although this has been true so far for earthquakes in the historical record, abundant evidence suggests that earthquakes, combined with other factors such as political instability or extended warfare, can significantly change a society and even the course of history. One purpose of this book has been to show that this has in fact happened.

It is important to remember, after all, that earthquakes do not occur in a vacuum but always in relation to other events. Although earthquakes seem to happen randomly—and certainly their triggers, in most cases, are entirely unrelated to human affairs—this does not mean their physical effects are separate from the actions of society. Instead, they are part of the natural landscape, like storms, droughts, volcanic eruptions, and plague, and are as likely to occur at times of instability as in times of peace and prosperity. Archaeologists, in their quest for single explanations and unifying theories, are often content to find evidence of war and then to assume that all the destruction in a given layer at an archaeological dig is a result of one cause. I argue that, at least in areas where geological and geophysical data show that earthquakes are a known recurrent hazard, we must always consider earthquakes in episodes of sudden total destruction.

This contrast in how archaeologists and geologists approach the past reflects a difference in our philosophical constitutions. As discussed in chapter 1, both our fields are observational, historical sciences and cannot be ruled by a strict falsifiability doctrine such as Karl Popper's. Although we share that limitation, however, archaeologists and geologists turn to different philosophical assumptions. Archaeologists and historians such as Tainter, Diamond, and Drews are "Toynbeeans," their thinking guided by Toynbee's assertion that the causes for societal collapse must lie within the society itself. In contrast, geologists are "Durantians," predisposed to think, like Durant, that "society exists by geological consent, subject to change without notice" (Byrne 1988). The truth is certainly a combination of these underlying beliefs, with geological and other environmental factors inextricably linked to the health of a society.

A well-researched, global, seismic hazard map is essential for publicizing the real risk of earthquakes around the world, and earthquake specialists are making much progress toward that goal. However, progress is necessarily slow, largely because we must wait for earthquakes to happen. With every new earthquake, our observation window becomes wider, and both the recurrence patterns of large earthquakes and the types of damage they cause become clearer. But it seems a shame to do nothing but wait for future earthquakes, when the other end of the observation window—the past—lies buried beneath our feet, or exposed in tourist-attracting ruins and ancient monuments, or obscured and encoded in our religious and secular histories.

I find it ironic that, to uncover evidence of past earthquakes, we must overcome the same dismissive attitude toward earthquakes that we are hoping to eventually change with the evidence we seek. Perhaps, by writing this book, I can help to open the eyes of both the archaeological community and the public to the facts I know to be true: the earth beneath our feet is not always steady, and its past cataclysms can be one key to understanding not only our prehistory but our future as well.

GLOSSARY

Acropolis. The high point of an ancient Greek city. The best known is the Acropolis of Athens, a large hill, which today has remnants of ancient temples and buildings.

aftershock. A smaller earthquake that follows a larger one, occurring on or near the part of the fault plane that slipped during the main shock. Aftershocks can occur for months or years following a large earthquake, generally decreasing in size and frequency as time passes.

Apocalypse. The battle at the end of the world, as described in the Book of Revelation in the Bible. Sometimes also used to describe any religiously charged major societal upheaval.

archaeology. The scientific study of ancient people through the examination of material remains and physical data, for example, skeletons, buildings, and artifacts.

archaeoseismology. An emerging field with significant impact on the understanding of earthquake science, archaeological studies, and historical studies. Also referred to as "earthquakes in archaeology."

Armageddon. An ancient city in Israel, also known as Megiddo.

artifact. A human-made object of archaeological interest.

Bronze Age. Period between the Stone Age and the Iron Age, dating between ca. 3000 BC and 1200 BC, distinguished by the use of bronze as a material for tools and weapons.

Catastrophism. The geologic theory that the earth's surface was shaped mostly by sudden upheavals, separated by long periods with little or no alteration. The rise of *Uniformitarianism* in the late eighteenth century threw Catastrophism into disfavor. However, most modern geologists agree that both gradual processes and catastrophes are important in the geologic record. See **Neo-catastrophism.**

Church of the Holy Sepulchre. A famous chapel in Jerusalem, built in the early fourth century. Revered by Christians as the site of the death, burial, and resurrection of Jesus Christ.

Colosseum, Rome. A famous oval amphitheater in Rome, built in the late first century AD. Most of it remains standing today, but part of the outer shell was destroyed by an earthquake.

Colossus of Rhodes. A statue of the Greek god Helios that was built at the entrance to the harbor of Rhodes and was one of the Seven Wonders of the Ancient World.

crust. The outermost layer of the earth's rigid **lithosphere.**

Dead Sea scrolls. A collection of manuscripts, many fragmentary, mostly dating to the first century AD and earlier, found in caves near the Dead Sea. They fall into several categories, including parts of Old Testament texts, a collection of community rules thought by many to refer to the community at Qumran, and miscellaneous unrelated writings. Collectively, they constitute one of the most important archaeological finds in the region, illuminating the formative period of both Christianity and modern Judaism.

displacement. Permanent shifting of the ground surface during an earthquake.

earthquake science, archaeological studies, and historical studies. Also referred to as "earthquakes in archaeology."

earthquake storm. A series of several major earthquakes that occur in one region within a few months or years of one another. The earthquakes may be all on one fault, or may be on several different faults. It is hypothesized that each earthquake redistributes stress in the region, making other faults, or other sections of the same fault, more likely to fail.

earthquake. Shaking of the earth's surface caused by volcanic activity, meteorite impacts, human activity, or, most commonly, tectonic motion on faults. See **tectonic earthquakes**.

elastic rebound. A theory explaining how the slow, steady motion of **plate tectonics** can result in the episodic and violent motion seen in earthquakes. As **plates** move past one another, the **faults** at their boundaries often become stuck, so that the plate motion warps the rocks in the vicinity of the fault. Energy is stored in the strained rocks until the energy exceeds the strength of the fault; then sides of the fault slip past one another, and the rocks rebound into their unstressed shape. By this mechanism, all the energy stored in the rocks over years of plate motion is released in a few seconds, in an earthquake.

End of the Bronze Age. A transition that occurred around 1200 BC, and was marked by the destruction of many important sites in the Eastern Mediterranean and Aegean region.

enosichthon. "Earth shaker." See **Poseidon.**

epicenter. The point on the earth's surface directly above the **focus** of an earthquake, used to mark the location of an earthquake on a seismicity map. The actual focus may be many kilometers below the surface.

Essenes. An ancient religious sect, thought by many to be the inhabitants of Qumran on the Dead Sea, the community associated with the Dead Sea Scrolls.

Falsifiability. A doctrine holding that a theory is scientific only if experiments can be conducted to eliminate the theory if it is false.

fault scarp. A linear ridge on the earth's surface caused by vertical displacement along a normal or reverse fault.

fault. A plane of weakness in the earth's **crust** that accommodates motion between plates or smaller blocks of crust. Earthquakes occur when energy stored in the rocks around the fault is released, with one side of the fault slipping suddenly past the other. In any given earthquake, only a portion of the fault will slip, and that slip may or may not break the rocks at the surface. Many faults are undetectable at the earth's surface until seismographs record earthquakes on them.

focus. The point on a fault where the rocks first begin to slip past one another. An earthquake focus can be anywhere from a few kilometers to hundreds of kilometers beneath the earth's surface. Seismographs are used to locate earthquake foci.

foreshock. Small earthquakes that can occur as precursors to larger shocks. Unfortunately, foreshocks are generally recognized as such only in retrospect, after the larger event occurs.

frequency-magnitude relation. A simple statistical rule that seems to govern the occurrence of small and large earthquakes: $\log(N) = A - BM$, where N is the cumulative number of events, B is a constant that depends on the overall seismic risk in the region, and A is a constant related to the size of the area sampled and the length of time for which the data have been collected. Simply stated, the equation implies that if scientists accumulate a record of small earthquakes in a region, they can estimate how often, *on average*, larger earthquakes will strike.

geophysics. The study of the physics of the earth and its atmosphere, with many sub-disciplines, including seismology.

global positioning system (GPS). A radio navigation and mapping system using a collection of satellites in geostationary orbits. By installing permanent markers and tracking their motion over time with GPS receivers, geophysicists can track the slow motion of the earth's tectonic plates and measure displacements that occur in earthquakes.

Harvard Syriac 99. An early fifth-century Syriac letter, credited to Cyril, Bishop of Jerusalem, containing specific details of the 363 AD earthquake that struck the Holy Land. Sebastian P. Brock discovered the rewritten copy, dated 1899, in the Houghton Library at Harvard University, and translated and published it in 1977.

Helots. The lowest class of serfs in ancient Sparta, Greece.

Hittites. Early rulers of most of Anatolia, now modern Turkey. Their ancient capital was Hatushash, around 1700–1200 BC.

Homer. Ancient Greek blind poet ascribed with the composition of two epic poems, the *Iliad* and the *Odyssey*. The works were transmitted orally for centuries before being recorded in writing, leading some scholars to question whether Homer really existed or was the amalgamation of several anonymous authors.

human sacrifice. The ritual killing of human beings to appease the gods, practiced in certain cultures and religions. Some cultures believed that an angry god caused earthquakes, and that human sacrifice might prevent further disasters.

Hyksos. Foreigners who invaded and ruled Egypt around 1700 BC.

intensity. A measure of earthquake size that reflects the effects on humans and human construction. The most commonly used scale for measuring earthquake intensity in the United States is the Modified Mercalli Scale; in Europe, the Rossi-Forel scale is used.

intraplate earthquake. An earthquake that occurs far from plate boundaries, in the usually more stable plate interior. These earthquakes tend to have very long recurrence intervals and are therefore poorly understood. The best known intraplate earthquakes in the United States are the New Madrid, Missouri, earthquakes of 1811 and 1812 .

isoseismal map. A map showing contours of equal seismic intensity in an earthquake. Sites within a given contour suffered similar damage.

keystone. The central voussoir in an arch, often larger or more prominent than the other voussoirs. See **voussoir.**

lighthouse at Alexandria. Also known as the Pharos of Alexandria, one of the Seven Wonders of the Ancient World, which suffered repeated earthquake damage and was completely destroyed by an earthquake in the fourteenth century.

Linear B. A written form of early Greek developed in the second millennium BC, found on clay tablets at several ancient sites, including Knossos and Mycenae.

liquefaction. A loss of cohesion that occurs when water-saturated soil or sediment is shaken by an earthquake. The soil loses its strength and can flow out from under the foundations of buildings or bridges, and can even erupt at the surface in transient sand blows or mud "volcanoes."

lithosphere. The rigid layer at the earth's surface, made up of tectonic plates, that floats on the softer layers below. The ocean floors and continents are all part of the lithosphere.

magnitude, local. A measure of earthquake size first defined by Charles Richter in 1935, and commonly known as the Richter magnitude. The local magnitude is determined by the amplitude of waves recorded on a Wood-Anderson seismometer situated 100 kilometers from the epicenter of the earthquake. Like all magnitude scales, the local magnitude scale is logarithmic, meaning that each unit increase in magnitude represents a tenfold increase in amplitude, and is also open-ended, meaning that earthquakes can be arbitrarily large or small, although very large earthquakes are poorly differentiated by local magnitude. Local magnitude is designated by the symbol M_L.

magnitude, moment. A measure of earthquake size based on the seismic moment of the earthquake, as determined by analyzing seismograms and field data. The seismic moment reflects the size of the fault rupture, the strength of the rocks, and the amount of displacement between the two sides of the rupture surface. The moment magnitude scale converts the seismic moment into a logarithmic magnitude. Unlike the local magnitude, the moment magnitude provides a good description of earthquake size even for very large or very small quakes. Moment magnitude is designated by the symbol M_w.

Mapuche Indians. The inhabitants of the Pacific coastal town of Lago Budi, Chile. Their culture attributes earthquakes to a wicked snake, Cai Cai, who can be appeased by human sacrifice.

Namazu. In Japanese mythology, a giant catfish responsible for earthquakes.

Neo-catastrophism. A term used to describe the resurgent interest in natural disasters to explain prehistoric events. Phenomena as disparate as severe drought, climate variations, volcanic eruptions, meteorite strikes, and earthquakes are lumped together under this designation, which is commonly contrasted with the scientific principle of Uniformitarianism. This comparison is not strictly appropriate, however. Uniformitarianism simply states that the physical laws governing earth processes today are the same as those that operated in the past. Neo-catastrophism more correctly should be contrasted with gradualism, which holds that all changes in the past resulted from the action of long-term processes that continue to act today. Neo-catastrophism relies on scientifically explainable phenomena, whereas the older philosophy of Catastrophism sometimes invoked biblical cataclysms like the Flood.

normal fault. A type of fault that commonly occurs at divergent boundaries, where rocks on one side of the fault, called the hanging block, slide down the inclined fault surface. These types of faults tend to produce topographic depressions.

Occam's razor. An axiom applied in science and philosophy that hypotheses should not be multiplied unnecessarily. Attributed to the early-fourteenth-century English philosopher, William of Ockham.

Optimism. A philosophy articulated in the early eighteenth century by philosophers like Gottfried Wilhelm von Leibniz (1646–1716) and Alexander Pope (1688–1744), arguing that every event and process worked toward the general good of humankind, yielding the best of all possible worlds.

oral tradition. Oral transmission of cultural and historical traditions from one generation to another, without a written system. In the Greek culture, epic poems, which show mythological elements, started as oral tradition and were later recorded in writing.

palynology. The scientific examination of pollen grains preserved in ancient rocks, soils, and sediments to determine past vegetation, and to understand ancient environmental and climatic conditions.

Pebble Man. Name given to the people, thought to be members of *Homo erectus,* who created the oldest known human construction, the pebble floor at **Ubediyeh.**

plate boundary. The narrow region between two lithospheric plates, accommodating motion between them. Plate boundaries are the source of most earthquakes, and can be convergent (where two plates collide), divergent (where two plates move apart), or transform (where plates move past one another laterally).

plate tectonics. A broad theory that explains many features of the earth's surface, including the locations and formation of the continents and oceans, the formation of volcanoes, and the distribution of earthquakes. According to plate tectonics, the earth's outer layer, the lithosphere, is divided into broad, stable regions called plates, which move past one another at speeds of a few centimeters per year. The plates float on the softer layer beneath the lithosphere, called the asthenosphere. New lithosphere is created at divergent boundaries, mostly at mid-oceanic ridges. Older lithosphere is recycled into the earth's interior at convergent boundaries, or subduction zones, where one plate dives beneath another, creating a deep trench on one side of the boundary and volcanoes on the other.

plate. A stable and relatively earthquake-free region of the lithosphere that moves over the softer layers below, causing earthquakes at its boundaries.

Poseidon. According to ancient Greek mythology and religion, Poseidon is the Sea god and is responsible for earthquakes; he is mentioned in the works of Homer and Hesiod as "enosichthon," the shaker of the earth.

prehistory. The period before written records, as revealed by archaeology, geology, and mythology.

p-wave. See **seismic wave.**

Qumran. An ancient ruin situated under the limestone cliffs of the northwest shore of the Dead Sea. It sits on a very active earthquake belt, which is today in a tectonic quiet period. Some of the Dead Sea Scrolls were discovered in caves nearby, and many scholars believe the scrolls were associated with the site's inhabitants.

radiocarbon dating. A radiometric technique that uses radioactive isotope carbon-14 to reconstruct the age of organic materials contained within it.

Ramesseum. The burial temple of Ramses II, ruler of Egypt for sixty-six years, from 1278 to 1212 BC.

recurrence interval. The average elapsed time between earthquakes of a given magnitude, either for a particular fault or for a region. This is a purely statistical measure, as the actual interval between earthquakes varies greatly. The recurrence interval is useful, however, for assessing seismic hazard, developing seismic building codes, and analyzing cost-benefits of seismic retrofitting of existing structures.

reverse fault. A type of fault that commonly occurs at convergent boundaries, where rocks on one side of the fault, called the hanging block, move upward over the inclined fault surface. These faults, sometimes also called *thrust faults*, generally produce hills or mountains.

Richter Scale. An open-ended, logarithmic scale developed in 1935 by the American geophysicist Charles Francis Richter to describe what is now known as the local magnitude of an earthquake. See **magnitude, local.**

rift. A type of divergent plate boundary. A rift usually refers to an incipient divergent boundary that splits a continent, making a deep, volcanically active valley, like the East African Rift Valley. As the plates diverge further, the rift valley deepens, eventually flooding and becoming an ocean basin.

rupture. The breaking and slipping of a section of a fault in an earthquake. Although earthquakes begin at a point called a *focus*, the motion propagates outward along the fault plane as the quake progresses, and eventually a finite area of the fault moves while the rest remains intact. The size of the rupture area is a key determinant of the earthquake's magnitude.

San Andreas Fault. A major transform fault, extending more than 1,200 kilometers through coastal California, outlining the boundary between the North American Plate and the Pacific Plate.

Sea Peoples. The name applied to a group of unidentified aggressors credited with destroying many sites at the end of the Bronze Age in the Mediterranean region.

seiche. A standing wave that sloshes the water of a lake or small sea, the result of either an earthquake or wind patterns.

seismic wave. A form of energy that propagates outward from the slipping fault plane and is responsible for the shaking during an earthquake. There are many types of seismic waves, including *p-waves*, *s-waves*, and various types of surface waves. In p-waves, "p" stands for *primae*, or "first," because p-waves are the fastest-traveling waves and thus the first to arrive at seismographs. They are nothing more than sound waves traveling through the earth, alternately compressing and dilating the rocks they pass through. S-waves ("s" standing for *secundae*) are slower than p-waves but cause more damage. They are shear waves, which means they shake the earth "sideways," perpendicular to the direction they travel. Unlike p-waves, which can travel through any material, s-waves cannot pass through fluids. This useful fact allows seismologists to map the earth's liquid outer core.

seismicity. Earthquake activity in a particular area.

seismite. Layers of chaotically mixed sediments caused by underwater landslides during earthquakes and found in sea-floor or lake-bed deposits. Often, the mixed layers and the undisturbed layers deposited over them contain organic material that can be used to establish radiocarbon dates for the earthquakes.

seismograph. An instrument designed to record the shaking that occurs during an earthquake. Seismographs can be designed to record many types of seismic information, including arrival time of waves, maximum amplitude of strong earth motion, and direction of shaking. Arrays of seismographs are used to pinpoint the location of an earthquake's focus, as well as to analyze the detailed characteristics of its source.

seismometer. See ***seismograph***.

Shanidar Cave. A cave site with Neanderthal remains, located in the mountains of northeast Iraq, excavated by Ralph Solecki.

shear zone. The broad band of deformation that characterizes most major transform boundaries. A given transform boundary generally has several major faults and many smaller ones, most roughly parallel to one another, and all accommodating motion between the plates. The San Andreas plate boundary, for instance, includes the Hayward, San Jacinto, and Imperial faults, and many others besides the famous San Andreas Fault, all of which accommodate part of the motion between the North American and Pacific plates.

slip. The motion of two sides of a fault past each other.

societal collapse. The major breakdown of a society or civilization because of human or natural factors.

stratigraphy. The study of layers in the earth, for either geological or archaeological interpretation. Typically, not every distinct layer in a stratigraphic section can be dated, but those layers for which a date can be determined are used to constrain the dates for layers above and below them. See **superposition**.

stress. The force per unit area acting on a plane within a body. Six values are required to characterize completely the stress at a point: three normal components and three shear components.

strike. The orientation of a fault trace on the earth's surface. The term *strike-slip* is often used to describe transform faults, because the direction of motion, or slip, is parallel to the fault's strike.

subduction. The process by which one tectonic plate descends into the earth's interior beneath another plate at a convergent boundary. The descending plate is heated in the earth's interior, driving water and other volatile compounds out of the rock and causing magma to form above the plate. This magma rises through the overriding plate to form the volcanic mountain ranges that are typical of subduction zones. Where the descending plate bends down to plunge into the earth, a deep oceanic trench forms. The Marianas Trench, the deepest on earth, is a subduction feature.

subsidence. A lowering of the land's surface that can be caused by earthquakes, damaging buildings and affecting drainage patterns of rivers or streams.

superposition. The most basic principle of stratigraphy, which holds that younger layers of sediment or habitation overlie older layers, provided no subsequent event has perturbed them.

s-wave. see **seismic wave.**

trace. The line a fault plane makes where it intersects the surface. The trace of a fault at the surface is not always obvious unless recent motion on the fault has been significant.

transform fault. A type of fault that commonly occurs at transform boundaries, where rocks on one side of the fault move horizontally past rocks on the other side.

trenching. An investigative technique used to examine multiple layers of stratigraphy along a linear course. In archaeology, trenching is used to explore a potential excavation site in hopes of pinpointing particularly productive areas. In geology, trenching is typically used to cut across an active fault to reveal its seismic history. Layers that were broken in previous earthquakes become buried by sediment over time, making it possible to establish the number of times a fault has been reactivated in a given period.

tsunami. An ocean wave triggered by vertical movement of the sea floor in an earthquake. The sudden displacement of the sea floor pulls a large volume of water with it, momentarily shifting sea level in a broad region above the seafloor disturbance. The surface disturbance radiates outward across the ocean as a wave many kilometers wide but only a few centimeters or meters high. When the wave approaches land, however, its height increases, causing enormous damage when it makes landfall. A tsunami can travel across a whole ocean, striking distant shores hours after the earthquake that generated it.

Ubediyeh. The site of the oldest known human construction, south of the Sea of Galilee on the banks of the Jordan River. The construction is a floor made of pebbles and discarded bones pressed into the muddy river bank, now tilted sixty degrees from horizontal by the accumulated deformation caused by earthquakes. See also **Pebble Man.**

Via Maris. A major historic trade route linking Egypt with Syria, Anatolia, and Mesopotamia.

voussoir. Any of the wedge-shaped stones supporting the opening of an arched doorway or vaulted ceiling. These stones often slip downward when an earthquake widens the doorway or vault. If the change in width is small enough, the voussoirs may slip only partway, locking the arch in its new, wider position.

REFERENCES

Agnew, D. C.

2002 History of seismology. In: (W. H. K. Lee, H. Kanamori, P. C. Jenning, and C. Kisslinger, Eds.) *International Handbook of Earthquake and Engineering Seismology, Part A,* 3–12. London and San Diego, CA: Academic Press.

Al-Homoud, A. S.

2000 Geologic hazards of an embankment dam constructed across a major, active plate boundary fault. *Geoscience* 6 (4), 353–382.

Allegro, J.

1964 *The Dead Sea Scrolls: A Reappraisal.* London and New York: Penguin Books.

Allen, C. R.

1975 Geological criteria for evaluating seismicity. *Geological Society of America Bulletin* 86, 1041–1057.

Allen, S. H.

1996 Calvert's heirs claim Schliemann treasure. *Archaeology* 49 (1), 22. Available at: http://www.archaeology.org/9601/index.html. Accessed May 10, 2007.

1999 *Finding the Walls of Troy: Frank Calvert and Heinrich Schliemann at Hisarlik.* Berkeley: University of California Press.

Alsop, J.

1981 A historical perspective. *National Geographic* 159 (2), 223.

Ambraseys, N. N.

1970 Some characteristic features of the Anatolian Fault Zone. *Technophysics* 9, 143–165.

1971 Value of historical records of earthquakes. *Nature* 232, 375–379.

Ambraseys, N. N., and C. F. Finkel

1987 Seismicity of Turkey and neighbouring regions, 1899–1915. *Annales Geophysicae* 5, 701–725.

1988 The Anatolian Earthquake of 17 August 1668. In: (W. H. K. Lee, H. Meyers, and K. Shimazaki, Eds.) *Historical Seismograms and Earthquakes of the World,* 173–180. New York: Academic Press.

Ambraseys, N. N., and J. A. Jackson
1998 Faulting associated with historical and recent earthquakes in the Eastern Mediterranean region. *Geophysical Journal International* 133, 390–406.
Ambraseys, N. N., and I. Karcz
1992 The earthquake of 1546 in the Holy Land. *Terra Nova* 4 (2), 254–263.
Ambraseys, N. N., and C. P. Melville
1982 *A History of Persian Earthquakes*. Cambridge: Cambridge University Press.
Ambraseys, N. N., C. P. Melville, and R. D. Adams
1994 *The Seismicity of Egypt, Arabia and the Red Sea: A Historical Review*. Cambridge: Cambridge University Press.
Amiran, D. H. K., E. Arieh, and T. Turcotte
1994 Earthquakes in Israel and adjacent areas: Macroseismic observations since 100 B.C.E. *Israel Exploration Journal* 44, 260–305.
Angulo, J.
1996 *Teotihuacan: City of the Gods*. (English ed.) Mexico: Monclem Ediciones.
Aristotle
1952 *Meteorologica*. (H. D. P. Lee, Ed. and Trans.) Loeb Classical Library. London and Cambridge, MA: Harvard University Press.
Armijo, R., A. Deschamps, and J. P. Poirier
1986 Carte Sismotectonique: Europe et Bassin Mediterraneen. Paris: Institut de Physique du Globe de Paris.
Armijo, R., H. Lyon-Caen, and D. Papanastassiou
1991 A possible normal-fault rupture for the 464 B.C. Sparta earthquake. *Nature*, 351, 123–125.
Åström, P.
1968 The destruction of Midea. *Atti e memorie del lo Congresso Internationale di Micenologia. Incunabula Graeca* 25, 54–57.
Åström, P., and K. Demakopoulou
1996 Signs of an earthquake at Midea? In: (S. Stiros and R. E. Jones, Eds.) *Archaeoseismology*, 37–40. Fitch Laboratory Occasional Paper No. 7, Athens. Exeter: Short Run Press.
Aurell, K. E.
1923 Quoted in B. S. Moore and wife, *The Japanese Disaster or The World's Greatest Earthquake: Together with Missionary Travels and Experiences*. Los Angeles: Giles.
Avni, R., D. Bowman, A. Shapira, and A. Nur
2002 Erroneous interpretation of historical documents related to the epicenter of the 1927 Jericho earthquake in the Holy Land. *Journal of Seismology* 6, 469–476.
Bar-Yosef, O.
1993 Ubeidiya. In: (E. Stern, Ed.) *The New Encyclopedia of Archaeological Excavations in the Holy Land*, 1487–1488. Carta, Jerusalem: Israel Exploration Society.

Begley, S.
1995 Lessons of Kobe. With P. McKillio and S. Strasser. *Newsweek*, 30 January 1995, 24–29.
Benedetti, L., R. Finkel, D. Papanastassiou, G. King, R. Armijo, F. Ryerson, D. Farber and F. Flerit
2002 Post-glacial slip history of the Sparta fault (Greece) determined by ^{36}Cl cosmogenic dating: Evidence for non-periodic earthquakes. *Geophysical Research Letters* 29, 8.
Ben-Menahem, A., A. Nur, and M. Vered
1976 Tectonics, seismicity and structure of the Afro-Eurasian junction: The breaking of an incoherent plate. *Physics of the Earth and Planetary Interiors* 12, 1–50.
Besterman, T.
1969 *Voltaire*. Chicago: University of Chicago Press.
Bienkowski, P.
1990 Jericho was destroyed in the Middle Bronze Age, not the Late Bronze Age. *Biblical Archaeology Review*, 45–69.
Blegen, C. W.
1963 *Troy and the Trojans*. New York: Praeger.
Blegen, C., and M. Rawson
1966 *The Palace of Nestor at Pylos in Western Messenia*. Vol. 1, *The Buildings and Their Contents*. Princeton, NJ: Princeton University Press.
Blegen, C. W., J. L. Caskey, and M. Rawson
1953 *Troy III: The Sixth Settlement*. Princeton, NJ: Princeton University Press.
Blegen, C. W., C. Boulter, J. L. Caskey, and M. Rawson
1958 *Troy IV: Settlements VIIa, VIIb and VIII*. Princeton, NJ: Princeton University Press.
Bolívar, S.
1983 [1812] *Cartagena Manifesto*. In: *Simon Bolívar: The Hope of the Universe*, 69–70. Paris: United Nations Educational Scientific and Cultural Organizations (UNESCO).
Bolt, B. A.
1993 *Earthquakes—Newly Revised and Expanded*. New York: Freeman.
Boschi, E., A. Caserta, C. Conti, M. Di Bona, R. Funiciello, L. Malagnini, F. Marra, G. Martines, A. Rovelli, and S. Salvi
1995 Resonance of subsurface sediments: An unforeseen complication for designers of Roman columns. *Bulletin of the Seismological Society of America* 85 (1), 320–324.
British Historical Society of Lisbon
1990 Lisbon recalled: All Saints Day, 1 November 1755. *Terramoto de 1755. Testemunhos Britanicos*, British account Lisboa, 670–672.
Brock, S. P.
1976 The rebuilding of the Temple under Julian: A new source. *Palestine Exploration Quarterly* 108, 103–107.

1977 A letter attributed to Cyril of Jerusalem on the rebuilding of the Temple. *Bulletin of the American Schools of Oriental and African Studies* 40, 267–286.

Browning, I.

1982 *Jerash and the Decapolis*. London: Chatto & Windus.

1989 *Petra*. London: Chatto & Windus.

Brumbaugh, D. S.

1999 *Earthquakes: Science and Society*. Upper Saddle River, NJ: Prentice Hall.

Bruins, H. J., and J. van der Plicht

1996 The Exodus enigma. *Nature* 382, 213–214.

Bullen, K. E.

1963 *An Introduction to the Theory of Seismology*. Cambridge: Cambridge University Press.

Bullen, K. E., and B. A. Bolt

1985 *An Introduction to the Theory of Seismology*. (4th ed.) Cambridge: Cambridge University Press.

Byrne, R.

1988 *1,911 Best Things Anybody Ever Said*. New York: Fawcett Columbine.

Cahill, D. P.

2002 The virgin and the Inca: An Incaic procession in the city of Cuzco in 1692. *Ethnohistory* 49 (3), 611–649. Duke University Press.

Calaprice, A.

2000 *The Expanded Quotable Einstein*. Princeton, NJ: Princeton University Press.

Catling, H. W.

1981 Archaeology in Greece, 1980–81. *Archaeological Reports for 1980–81*, 3–48.

Chamoto, S.

1984 *Senso to Janarizumu*. Tokyo: San-ichishobou. (Quoted by Ryouko Hatari, Yukiko Imai, and Ewa Watanabe. In *The Wild Rumor: Understanding the Anti-Korean Riots of 1923*). Available at: http://www.tsujiru.net/compass/compass_1996/reg/hatari_imai_watanabe.htm. Accessed May 9, 2007.

Cline, Eric

2004 *Jerusalem Besieged: From Ancient Canaan to Modern Israel*. Ann Arbor: University of Michigan Press.

Coe, M. D.

1962 *Mexico: From the Olmecs to the Aztecs*. New York: Thames and Hudson.

1998 *Archaeological Mexico*. Chico, CA: Moon Travel Handbooks.

Crown, A. D., and L. Cansdale

1994 Qumran: Was it an Essene settlement? *Biblical Archaeology Review* 20 (5), 24–35, 73–74, 76–78.

Dakoronia, P.

1996 Earthquakes of the Late Helladic III period (12th century BC) at Kynos (Livanates, Central Greece). In: (S. Stiros and R. E Jones, Eds.) *Archaeoseismology*, 41–44. Exeter: Short Run Press.

Darwin, C.
1982 [1845] Journal of researches into the natural history and geology of the countries visited during the voyage of *HMS Beagle* round the world (2nd Ed.). In: (G. Y. Craig and E. J. Jones, Comps.) *A Geological Miscellany*, 33–34. Oxford: Orbital.
Davies, G.
1986 *Megiddo*. Cambridge: Lutterworth.
Demakopoulou, K.
1998 Stone vases from Midea. In: (E. Cline and D. Harris-Cline, Eds.) *The Aegean and the Orient in the Second Millennium. Proceedings of the 50th Anniversary Symposium Cincinnati, 18–20 April 1997*, 221–227. Aegaeum 18. Liege: Université de Liege.
de Vaux, R.
1973 *Archaeology and the Dead Sea Scrolls*. London: Oxford University Press.
Dewey, J. R. and A. M. C. Sengör
1979 Aegean and surrounding regions: Complex multiplate and continuum tectonics in a convergent zone. *Geological Society of America Bulletin* 90, 84–92.
Diamond, J.
2005 *Collapse: How Societies Choose to Fail or Succeed*. New York: Viking
Diodorus Siculus
1994 *The Library of History Books II.35–IV.58*. (Trans. C. H. Oldfather; Ed. G. P. Goold) Vol. 2. Loeb Classical Library. Cambridge, MA: Harvard University Press.
Donceel , R., and P. Donceel-Voute
1994 *Methods of Investigation of the Dead Sea Scrolls and the Khirbet Qumran Site: Present Realities and Future Prospects*. (Ed. M. O. Wise, N. Golb, J. J. Collins and D. Pardee.) New York: Annals of the New York Academy of Sciences.
Drews, R.
1993 *The End of the Bronze Age: Changes in Warfare and the Catastrophe ca. 1200 B.C.* Princeton, NJ: Princeton University Press.
Earthquake Engineering Research Institute
2004 Preliminary observations on the Bam, Iran, earthquake of December 26, 2003. Earthquake Engineering Research Institute (EERI) Special Earthquake Report, April. Available at: http://www.eeri.org/lfe/iran_bam.html. Accessed May 9, 2007.
Ellenblum, R., S. Marco, A. Agnon, T. Rockwell, and A. Boas.
1998 Crusader castle torn apart by earthquake at dawn, 20 May 1202. *Geology* 26, 303–306.
EQE International
1995 *The January 17, 1995, Kobe Earthquake, EQE Summary Report*. Available at: www.absconsulting.com/CatastropheReports.html. Accessed May 9, 2007.

Evans, A.
1928 *Palace of Minos II*. London: Macmillan.
1964 *The Palace of Minos at Knossos*. New York: Biblo and Tannen.

Fast, P. A.
1997 *When the Earth Trembles*. Mukilteo, WA: Winepress.

Fiedel, S. J.
1994 *Prehistory of the Americas*. New York: Cambridge University Press.

Finkelstein, I., and D. Ussishkin
1993 Back to Megiddo. *Biblical Archaeology Review* 20, 28–43.

Fokaefs, A., and G. Papadopoulos
2004 Tsunamis in the area of Cyprus and the Levantine Sea. *European Geosciences Union* 6.

French, E. B.
1996 Evidence for an earthquake at Mycenae. In: (S. Stiros and R. E. Jones, Eds.) *Archaeoseismology*, 51–54. Exeter: Short Run Press.

Freund, R. A.
2004 *Secrets of the Cave of Letters*. Amherst, NY: Humanity Books.

Freund, R., Z. Garfunkel, I. Zak, M. Goldberg, T. Weissbrod, and B. Derin
1970 The shear along the Dead Sea Rift. *Philosophical Transactions of the Royal Society of London* A267, 107–130.

Frumkin, A. (Director), H. Eshel, S. Lisker, R. Porat, G. Danon, and R. Tsabar
2003 *Dead Sea Caves: Periodical Report #4*. Stanford University Rock Physics and Borehole Geophysics Project and Hebrew University of Jerusalem Cave Research Center.

Galanopoulos, A. G.
1963 On mapping of seismic activity in Greece. *Annali di Geofisica* 16, 37–100.
1968 On quantitative determination of earthquake risk. *Annali di Geofisica* 21, 193–206.
1973 Plate tectonics in the area of Greece as reflected in the deep focus seismicity. *Annali di Geofisica* 26, 85–105.

Ganse, A., and J. B. Nelson
1981 *Catalog of significant earthquakes 2000 B.C.–1979*. Boulder, CO: World Data Center A for Solid Earth Geophysics.

Garfunkel, Z.
1981 Internal structure of the Dead Sea leaky transform (rift) in relation to plate kinematics. *Tectonophysics* 80, 81–108.

Garfunkel, Z., I. Zak, and R. Freund
1981 Active faulting in the Dead Sea Rift. *Tectonophysics* 80, 1–26

Garstang, John, and J. B. E. Garstang
1940 *The Story of Jericho*. London: Hodder & Stoughton.

Gilbert, G. K.
1982 [1906] The investigation of the San Francisco earthquake. In: (G. Y. Craig and E. J. Jones, Comps.) *A Geological Miscellany*, 33–34. Oxford: Orbital. Originally published in *Popular Science Monthly* 69, 97.

Golb, N.
1995 *Who Wrote the Dead Sea Scrolls? The Search for the Secret of Qumran*. New York: Touchstone.

Gonen, R.
Megiddo in the Late Bronze Age—another reassessment. *Levant* 19, 83–100.
Grant, M., and J. Hazel
1993 *Who's Who in Classical Mythology*. New York: Oxford University Press.
Guidoboni, E., and A. Comastri,
1997 The large earthquake of 8 August 1303 in Crete: Seismic scenario and tsunami in the Mediterranean area. *Journal of Seismology* 1 (1), 55–72.
2003 *Catalogue of Earthquakes in the Mediterranean Area from the 11th to the 15th century*. Rome: Instituto Nazionale di Geofisica.
Guidoboni, E., A. Comastri, and G. Traina
1994 *Catalogue of Ancient Earthquakes in the Mediterranean Area up to the 10th Century*. Rome: Instituto Nazionale di Geofisica.
Hanfmann, G. M. A.
1951 The Bronze Age in the Near East: A review article, part I. *American Journal of Archaeology* 55, 355–365.
1952 The Bronze Age in the Near East: A review article, part II. *Journal of Archaeology* 56, 27–38.
Harding, G. L.
1959 *The Antiquities of Jordan*. London: Lutterworth.
Harvey, R.
2000 *Liberators: Latin America's Struggle for Independence, 1810–1830*. London: John Murray.
Heinsohn, G.
1998 The catastrophic emergence of civilization: The coming of blood sacrifice in the Bronze Age. In: (B. J. Peiser, T. Palmer, and M. E. Bailey, Eds.) *Natural Catastrophes during Bronze Age Civilisations*. British Archaeological Reports International Series 728. Oxford: Archaeopress.
Heylighen, F.
1997 Occam's razor. In: (F. Heylighen, C. Joslyn, and V. Turchin, Eds.) *Principia Cybernetica Web*. Brussels: Principia Cybernetica. Available at http://cleamc11.vub.ac.be/OCCAMRAZ.html. Accessed May 24, 2007.
Higgins, M., and R. Higgins
1996 *A Geological companion to Greece and the Aegean*. New York: Cornell University Press.
Homer
1990 *The Iliad* (Robert Fagles, Trans.). New York: Penguin Classics.
Iakovidis, S. E.
1986 Destruction horizons at late Bronze Age Mycenae. In: *Philia Epi eis Georgion E. Mylonan, v. A.*, 233–260. Athens: Library of the Archaeological Society of Athens.
International Millennium Publications (IMP)
1999 *2000 Years of Pilgrimage to the Holy Land* (D. Hadary-Salomon and D. Camiel, Eds.) Israel: Alfa Communication.
Jackson, J.
1993 Rates of active deformation in the eastern Mediterranean. In: (E. Boschi, E. Mankovani, and A. Morelli, Eds.) *Recent Evolution and Seismicity of the Mediterranean Region*, 53–64. Dordrecht: Kluwer Academic Press.

James, C. D., and S. Fatemi
2002 *Aftershocks*. Berkeley: University of California Press.
Jewish Telegraphic Agency
1927 Late Estimates on Palestine Quake Toll Place Dead at 690 and Injured at 3000. Jerusalem, July 14. Reprinted in Los Angeles Times, July 15, 1927, p. 2.
Joffe, S., and A. Garfunkel
1987 Plate kinematics of the circum Red Sea–A re-evaluation. *Tectonophysics* 141, 5–22.
Johnston, A. C.
1996 A wave in the earth. *Science* 274 (5288), 735.
Joseph, L. E.
The Growth of an Idea. New York: St Martin's.
Josephus, Flavius
1982 [AD 75] *The Jewish War* (G. Cornfield, Ed.). Grand Rapids, MI: Zondervan.
1991a [AD 75] *Wars of the Jews*. In: *The Complete Works of Josephus* (William Whiston, Trans.), chap. 19, 427–605. Grand Rapids, MI: Kregel.
1991b [AD 93] *Antiquities of the Jews*. In: *The Complete Works of Josephus*, bk. 15, chap. 5, 23–426. Grand Rapids, MI: Kregel.
Karnik, V.
1968 *Seismicity of the European Area*. Part 2. Prague: Academia Press.
Katayama, T.
1996 Lessons from the 1995 Great Hanshin Earthquake of Japan with emphasis on urban infrastructure systems. In: *Proceedings of Structural Engineering in Consideration of Economy, Environment and Energy: 15th Congress Report. (Congress of International Association for Bridge and Structural Engineering—IABSE; June 1996)*, 187–200. Copenhagen: IABSE. Available at http://www.greenbar.org/arcive/kataya/pub1-1.html. Accessed May 15, 2007.
Kendrick, T. D.
1956 *The Lisbon Earthquake*. London: Methuen.
Kempinski, A.
1993 *Megiddo: A City-State and Royal Centre in North Israel*. Munich: Verlag C. H. Beck; Tel-Aviv: Hakkibbutz Hameuchad.
Ken-Tor, R., A. Agnon, Y. Enzel, M. Stein, S. Marco, and J. F. W. Negendank
2001 High-resolution geological record of historic earthquakes in the Dead Sea basin. *JGR* 106 (B2), 2221–2234.
Kenyon, K. M.
1957 *Digging up Jericho*. New London: Praeger.
1979 *Archaeology in the Holy Land*. London: E. Benn.
Kenyon, K. M., and A. D. Tushingham
1953 Jericho gives up its secrets. *National Geographic* 104 (6), 853–870.
Kilian, K.
1996 Earthquakes and archaeological context at 13th century BC Tiryns. In: (S. Stiros and R. E. Jones, Eds.) *Archaeoseismology*, 63–68. Fitch Laboratory Occasional Paper No. 7, Athens. Exeter: Short Run Press.

King, G. C. P., R. S. Stein, and J. Lin
1994 Static stress changes and the triggering of earthquakes. *Bulletin of the Seismological Society of America* 84, 935–953.
Koto, B.
1893 On the cause of the great earthquake in central Japan, 1891. *Journal of College Science*, 5, 296–353. Imperial University, Japan.
Kovach, R., and B. McGuire
2003 *Phillip's Guide to Global Hazards*. London: Octopus.
Kozak, J. T., and C. D. James
1998 *Historical Depictions of the 1755 Lisbon Earthquake*, National Information Service for Earthquake Engineering, University of California, Berkeley. Available at: http://nisee.berkeley.edu/lisbon/. Accessed May 9, 2007.
Leigh, J. A. (Ed.)
1967 *Rousseau to Voltaire, 18 August 1756, correspondence complete de Jean Jacques Rousseau. Geneva, 1967* (R. Spang, Trans.) 4, 37–50.
Leonard, A., Jr., and E. H. Cline
1998 The Aegean pottery found at Megiddo: An appraisal and reanalysis. *Bulletin of the American Schools of Oriental Research* 309, 3–39.
Marcellinus, Ammianus
Histories 17.7.8.
Magness, J.
2002 *Archaeology of Qumran and the Dead Sea Scrolls*. Grand Rapids, MI: Eerdmans.
Malkawi, A. H., and A. S. Alawneh
2000 Paleoearthquake features as indicators of potential earthquake activities in the Karameh Dam site. *Natural Hazards* 22 (1), 1–16.
Manzanilla, L.
2003 The abandonment of Teotihuacan. In: (T. Inomata and R. W. Webb, Eds.) *The Archaeology of Settlement Abandonment in Middle America*, 91–101. Salt Lake City: University of Utah Press.
Maroukian, H., K. Gaki-Papanastassiou, and D. Papanastassiou
1996 Geomorphologic-seismotectonic observations in relation to the catastrophes at Mycenae. In: (S. Stiros and R. E Jones, Eds.) *Archaeoseismology*, 189–194. Fitch Laboratory Occasional Paper No. 7, Athens. Exeter: Short Run Press.
Martin, T. R.
1996 *Ancient Greece from Prehistoric to Hellenistic Times*. New Haven and London: Yale University Press.
Maxwell, K.
1995 *Pombal, Paradox of the Enlightenment*. New York: Cambridge University Press.
McKenzie, D. P.
1970 Plate tectonics of the Mediterranean region. *Nature* 226, 239–243.
1972 Active tectonics of the Mediterranean region. *Royal Astronomical Society Geophysical Journal* 30, 109–185.

Milik, J. T.
1959 *Ten Years of Discovery in the Wilderness of Judaea.* (J. Strugnell, Trans.) London: SCM Press.
Moczo, P., P. Labák, and A. Rovelli
1995 Effects of the lateral heterogeneity beneath Roman Colosseum on seismic ground motion. *Proceedings of the 10th European Conference on Earthquake Engineering.* Vienna: A. A. Balkema, Rotterdam.
Moore, B. S., and wife
1923 *The Japanese Disaster, or The World's Greatest Earthquake: Together with Missionary Travels and Experiences.* Los Angeles: Giles.
Morkot, R.
1996 *The Penguin Historical Atlas of Ancient Greece.* London and New York: Penguin Books.
Mylonas, G.
1962 Excavations at Mycenae. *Praktika Archaeologikis Etaireias,* 57–66.
1963 Excavations at Mycenae. *Praktika Archaeologikis Etaireias,* 99–106.
1966 *Mycenae and the Mycenaean Age.* Princeton, NJ: Princeton University Press.
1970 Excavations at Mycenae. *Praktika Archaeologikis Etaireias,* 118–124.
1971 Excavations at Mycenae. *Praktika Archaeologikis Etaireias,* 146–156.
1972 Excavations at Mycenae. *Praktika Archaeologikis Etaireias,* 114–126.
1973 Excavations at Mycenae. *Praktika Archaeologikis Etaireias,* 99–107.
1975 Excavations at Mycenae. *Praktika Archaeologikis Etaireias,* 153–161.
Mylonas-Shear, I.
1969 Mycenaean domestic architecture. Ph.D. thesis. University of Michigan, Ann Arbor.
1987 *The Panagia Houses at Mycenae.* Philadelphia: University Museum Press.
Nakajima, Y.
1973 *Kanto Daishinsai* (Kanto earthquake). Tokyo: Yuzankaku.
Narula, P. L.
1995 Geotechnical investigations of Killari (Latur) earthquake of 30th September, 1993: An overview. (A collection of papers presented at the workshop on "30th September, 1993 Killari Earthquake, Maharashtra" held at Hyderabad on 24th December, 1993.) (M. Ramakrishnan, B. S. R. Murty, K. D. Viswanatham, and L. Harendranath, Eds.). Geological Survey of India, Special Publications No. 27, pages 7–16. Calcutta, India: Geological Survey of India.
National Oceanic and Atmospheric Association (NOAA)
Memorandum EDS NGSDC-2, NOAA/National Geophysical Data Center, Boulder, Colorado, 53 pp. (plus appendices).
Nelson, H. H.
1913 *The Battle of Megiddo,* Ph.D. thesis, University of Chicago.
New American Bible
1986 New York: Catholic Book Publishing.
Newman, J. H.
1842 *Essay on Miracles.* In: *The Ecclesiastical History of M. l'abbé Fleury, from the Second Ecumenical Council to the end of the Fourth Century,*

translated with notes, and an Essay on the Miracles of the Period by the Rev. J. H. Newman. Oxford: John Henry Parker. London: Rivington.

Nur, A.

1991 And the walls came tumbling down. *New Scientist* 1776, 45–48.

1999 Earthquakes, Armageddon, and the Dead Sea Scrolls, *AAGP Bulletin* 82 (11), 2154–2155.

1998 The collapse of ancient societies by great earthquakes. In: (B. J. Peiser, T. Palmer, and M. Bailey, Eds.) *Natural Catastrophes during Bronze Age Civilisations: Archaeological, Geological, Astronomical, and Cultural Perspectives. Biblical Archaeology Review,* 140–147. International Series 728. Oxford: Archaeopress.

1998 The end of the Bronze Age by large earthquakes? In: (M. Bailey, T. Palmer, and B. J. Peiser, Eds.) *Natural Catastrophes during Bronze Age Civilizations,* 140–149. British Archaeological Reports. Oxford: Archaeopress.

Nur, A., and E. H. Cline

2000 Poseidon's horses: Plate tectonics and earthquake storms in the Late Bronze Age Aegean and Eastern Mediterranean. *Journal of Archaeological Science* 27 (1), 43–63.

Nur, A., and Z. Reches

1979 The Dead Sea rift: Geophysical, historical and archaeological evidence for strike slip motion. *Eos, American Geophysical Union Transactions,* 60, 18, 322.

Nur, A., and H. Ron

1996 The walls came tumbling down: Earthquake history of the Holy Land. In: (S. Stiros and R. E. Jones, Eds.) *Archaeoseismology,* 75–85. Exeter: Short Run Press.

1997 Earthquake! Inspiration for Armageddon. *Biblical Archaeology Review* 23 (4), 49–55.

1997 Armageddon's earthquakes. *International Geology Review* 39, 532–541.

Nur A., C. MacAskil, and H. Ron

1991 *The Walls Came Tumbling Down: Earthquake History of the Holy Land.* Video documentary, 57 min. Stanford University, Department of Geophysics.

Papazachos, B., and C. Papazachou

1997 *The Earthquakes of Greece.* Thessaloniki: P. Ziti.

Papazachos, B. C., P. E. Comninakis, G. F. Karakaisis, B. G. Karakostas, Ch. A. Papaioannou, C. B. Papazachos, and E. M. Scordilis

2000 A catalogue of earthquakes in Greece and surrounding area for the period 550 BC–1999, *Publication of the Geophysical Laboratory*, University of Thessaloniki.

Parker, T. S.

1998 An early church, perhaps the oldest in the world, found in Aqaba. *Near Eastern Archaeology* 61 (4), 254.

Pausanias

1918 Description of Greece [Periegesis Hellados] (W. H. S. Jones and D. Litt, Trans.). 4 vols. Vol. 1, Attica and Corinth. Cambridge, MA: Harvard University Press; London: Heinemann.

Pausanias (2nd Century AD).
1979 *Guide to Greece*. Vol. 2, *Southern Greece* (P. Levi, Trans.). London: Penguin Classics.
The People's Korea
1999 *Testimony by Survivor of Great Kanto Earthquake*. Available at: http://210.145.168.243/pk/122nd_issue/99120107.htm. Accessed February 6, 2004.
Philo
1929–1933 *Works* (F. H. Colson and G. H. Whitaker, Eds.) Vols. 1–5. New York: Putnam's.
1935–1942 *Works* (F. H. Colson and G. H. Whitaker, Eds.) Vols. 6–10. Cambridge MA: Harvard University Press.
Pixner, B.
1990 Church of the Apostles found on Mt. Zion. *Biblical Archaeology Review* 16 (3), 16–37, 60.
Pliny
1938 [AD 77] *Natural History* (H. Rackham, Ed. and Trans.) Cambridge, MA: Harvard University Press; London: Heinemann.
Plutarch
1960 *The Rise and Fall of Athens: Nine Greek Lives*. (Ian Scott-Kilvert, Trans.) London: Penguin Books.
Popper, K. R.
2002 [1959] *Logic of Scientific Discovery*. London and New York: Routledge.
1974 Replies to my critics. In: (P. A. Schilpp, Ed.) *The Philosophy of Karl Popper*, 2:961–1197. Carbondale: The Library of Living Philosophers, University of Southern Illinois.
Rapp, G., Jr.
1982 Earthquakes in the Troad. In: (G. Rapp and J. A. Gifford, Eds.) *Troy: The Archaeological Geology*, 43–58. Princeton, NJ: Princeton University Press.
1986 Assessing archaeological evidence for seismic catastrophes. *Geoarchaeology* 1 (4), 365–379.
R. B.
1694 *The General History of Earthquakes* [microform]. London: Printed for Nath. Crouch. Location: Green Library, Stanford University. Call number: MFILM 015:4 (Media Microtext Collection).
Reches, Z., and D. F. Hoexter
1981 Holocene seismic and tectonic activity in the Dead Sea area. *Tectonophysics* 80, 235–254.
Reches, Z., D. F. Hoexter, R. Freund, and Z. Garfunkel
1981 Holocene seismic and tectonic activity in the Dead Sea area, the Dead Sea fort. Selected papers of the International Symposium on the Dead Sea Rift. *Tectonophysics* 80, 235–254.
Rhys-Davies, J.
1994 *Killer Quakes of the Bible*, VHS. Directed by Stacey Foiles. Silver Springs, MD: Discovery Communications.
Richter, C. F.
1958 *Elementary Seismology*. San Francisco: Freeman.

Risk Management Solutions
1995 *What If the 1923 Earthquake Strikes Again? A Five-Prefecture Tokyo Region Scenario*. Menlo Park, CA: Risk Management Solutions.
Roberts, D.
2000 *The Life, Works, Travels of David Roberts. R. A.* (Text by Fabio Bourbon). New York: Rizzoli.
Romer, J., and E. Romer
1995 *The Seven Wonders of the World*. New York: Holt.
Rose, M.
1999 *Godzilla's Attacking Babylon!* Available at: http://www.archaeology.org/online/features/godzilla. Accessed May 9, 2007.
Roth, F.
1988 Modelling of stress patterns along the western part of the North Anatolian Fault Zone. *Tectonophysics* 152, 215–226.
Rothenberg, B., and Y. Aharoni
1960 *In the Footsteps of Kings and Rebels*. Tel Aviv: Masada.
Russell, K. W.
1980 The earthquake of May 19 A.D. 363. *Bulletin of the American Schools of Oriental Research* 238, 47–64.
Safrai, B.
1993 Recollections from 40 Years Ago: More Scrolls Lie Buried. *Biblical Archaeology Review*, 51–57.
Sakellarakis, Y., and E. Sapouna-Sakellaraki
1981 Drama of death in a Minoan temple. *National Geographic* 174, 205–222.
Salamon, A., A. Hofstetter, Z. Garfunkel, and H. Ron
1991 A new seismicity map of the Sinai subplate. Annual Meeting of the Israel Geological Society, Acco, Israel, April 24–25.
Sampson, A.
1996 Cases of earthquakes at Mycenaean and pre-Mycenaean Thebes. In: (S. Stiros and R. E Jones, Eds.) *Archaeoseismology*, 113–117. Fitch Laboratory Occasional Paper No. 7, Athens. Exeter: Short Run Press.
Schaeffer, C. F. A.
1948 *Stratigraphie Comparée et Chronologie de l'Asie Occidentale* (Comparative Stratigraphy and Chronology of Western Asia). London: Oxford University Press.
1968 Commentaries sur les letters et documents trouvés dans les bibliothèques privées d'Ugarit. *Ugaritica V* (Mission de Ras Shamra 16), 607–768. Paris: Geuthner.
Seidensticker, Edward
1983 *Low City, High City: Tokyo from Edo to the Earthquake*. New York: Knopf.
Seneca, Lucius Annaeus
65 AD *The Workes of Lucius Annaeus Seneca Both Morall and Naturall* (Thomas Lodge, Trans. [1614]). London: William Stansby.
Sharer, R. J.
1994 *The Ancient Maya*. Stanford: Stanford University Press.

Shklar, J. N.
1990 *The Faces of Injustice*. New Haven and London: Yale University Press.
Shapira, A.
1979 Re-determined magnitude of earthquakes in the Afro-Eurasian junction. *Israel Journal of Earth Science* 28, 107–109.
Shapira, A., R. Avni, and A. Nur
1993 Note: A New Estimate for the Epicenter of the Jericho Earthquake of 11 July 1927. *Israel Journal of Earth Science* 42, 93–96.
Shelmerdine, C. W.
1997 Review of Aegean prehistory VI: The palatial Bronze Age of the southern and central Greek mainland. *American Journal of Archaeology* 101, 537–585.
Shipton, G. M.
1939 *Notes on the Megiddo Pottery of Strata VI–XX*. Chicago: University of Chicago Press.
Sibylline Oracles.
1899 New edition revised after the Text of Ruch, Books I–XIV. Translated from the Greek into English blank verse by Milton S. Terry. New York: Eaton and Mains; Cincinnati: Curts and Jennings.
Sieberg, A.
1932 Erdbebengeographie. *Handbuch der Geophysik*, Band IV, Abschnitt VI, 527–1005. Berlin, Germany: Gebrüder Bornträger.
Smith, R. H.
1982 Preliminary report on the 1981 season of the Sydney/Wooster joint expedition to Pella (spring season). In *Jordan Department of Antiquities Annual*, 26:323-344.
Solecki, R. S.
1959 *Three Adult Neanderthal Skeletons from Shanidar Cave, Northern Iraq*, Annual Report, Smithsonian Institution, Pub. 4392, 603–635.
1971 *Shanidar: The First Flower People*. New York: Knopf.
Soren, D., and J. James
1988 *Kourion: The Search for a Lost Roman City*. New York, London, Toronto, and Sydney: Anchor Press, Doubleday.
Stein, R., A. Barka, and J. H. Dietrich
1997 Progressive failure on the North Anatolian fault since 1939 by earthquake stress triggering. *Geophysical Journal International* 128 (3), 594–604.
Stein, S., and E. A. Okal
2005 Speed and size of the Sumatra earthquake. *Nature* 434, 581–582.
Stern, E.
1993 The many masters of Dor. Part II, How bad was Ahab? *Biblical Archaeology Review* 19 (2), 18–36.
Stewart, A.
1993 A death at Dor. *Biblical Archaeology Review* 19, 30–36, 84.
Stucky, R. A.
1990 Schweizer Ausgrabungen in ez Zantur, Petra: Vorbericht der Kampagne 1988. In: *Annual of the Department of Antiquities of Jordan*, Vol. 34. Amman, Jordan.

Symeonoglou, S.
1987 *Kadmeia I*. Göteborg: Studies in Mediterranean Archaeology.
Tainter, J. A.
1988 *The Collapse of Complex Societies*. Cambridge: Cambridge University Press.
Taylor, J.
1993 *Petra*. London: Aurum.
Thompson, R. C.
1937 A New Record of an Assyrian Earthquake. *Iraq* 4, 186.
Thucydides
1910 [431 BC] *History of the Peloponnesian War*. (R. Crawley, Trans.; W. R. Connor, Ed.) London: Everyman.
Tierney, P.
1989 *The Highest Altar: Unveiling the Mystery of Human Sacrifice*. New York: Penguin Books.
Tokyo Metropolitan Government
1995 *Tokyo and Earthquakes*. Tokyo: Simul International (Trans.).
Toynbee, A. J.
1939 *Study of History*. Vol. 4. London: Oxford University Press.
Tsafrir, Y., and G. Foerster
1992 The dating of the "Earthquake of the Sabbatical Year" of 749 C.E. in Palestine. *Bulletin of the School of Oriental and African Studies, University of London*, 55 (2), 231–235.
Vaucaire, M.
1929 *Bolivar the Liberator* (Margaret Reed, Trans.). Boston: Houghton Mifflin; Cambridge, MA: Riverside.
Verdelis, N.
1956 Excavations at Pylos. *Archaiologikon Deltion, Chronika* 15B, 5–8.
Voltaire
1998 [1759] Candide and Zadig. English ed. (R. B. Boswell, Trans.). Koln: Konemann.
1911 [1756] Poem on the Lisbon disaster. In: *Selected Works of Voltaire* (Joseph McCabe, Ed. and Trans.). London: Watts.
Wace, A. J. B.
1951 Mycenae 1950. *Journal of Hellenic studies* 71, 254–257.
Wace, A. J. B., M. Holland, M. F. S. Hood, A. G. Woodhead, and J. M. Cook
1953 Mycenae, 1939–1952. *Annual of the British School at Athens* 48, 3–93.
Waley, P.
1991 *Tokyo: City of Stories*. New York: Weatherhill.
Walker, B. S.
1982 *Earthquake*. Alexandria, VA: Time-Life Books.
Webster, D.
2002 *The Fall of the Ancient Maya*. New York: Thames and Hudson.
Willis, B.
1928 Earthquakes in the Holy Land. *Bulletin of the Geological Society of America* 18, 73–103.

Wood, B.
1990 Dating Jericho's destruction: Bienkowski is wrong on all counts. *Biblical Archaeology Review*, 45–69.
Wood, M.
1996 *In Search of the Trojan War*. 2nd ed. Berkeley: University of California Press.
Wood, H. O., and F. Neumann
1931 Modified Mercalli intensity scale of 1931. *Bulletin of the Seismological Society of America* 4, 277–283.
Yadin, Y.
1971 *Bar-Kokhba*. London and Jerusalem: Weidenfeld and Nicolson.
1975 *Hazor: The Rediscovery of a Great Citadel of the Bible*. New York: Random House.
Zangger, E.
1991 Tiryns Unterstadt. In: (E. Penick and G. Wagner, Eds.) *Archaeometry '90. International Symposium on Archaeometry, Heidelberg, April 1991*, 831–840. Basel: Birkhäuser Verlag.
1993 *The Geoarchaeology of the Argolid*. Berlin: Gebrüder Mann Verlag.
1994 Landscape changes around Tiryns during the Bronze Age. *American Journal of Archaeology* 98, 189–212.
Zebrowski, E., Jr.
1997 *Perils of a Restless Planet: Scientific Perspectives on Natural Disasters*. Cambridge: Cambridge University Press.
Zias, J. E.
2000 The cemeteries of Qumran and Celibacy: Confusion laid to rest? *Dead Sea Discoveries* 7 (2), 220–253.

INDEX

Figures are indicated with *italic* type.